3 4028 05800 5696
HARRIS COUNTY PUBLIC LIBRARY
W9-DDU-219

ALBERT EINSTEIN

Recent Titles in Greenwood Biographies

Fidel Castro: A Biography
Thomas M. Leonard

Oprah Winfrey: A Biography
Helen S. Garson

Mark Twain: A Biography
Connie Ann Kirk

Jack Kerouac: A Biography
Michael J. Dittman

Mother Teresa: A Biography
Meg Greene

Jane Addams: A Biography
Robin K. Berson

Rachel Carson: A Biography
Arlene R. Quaratiello

Desmond Tutu: A Biography
Steven D. Gish

Marie Curie: A Biography
Marilyn Bailey Ogilvie

Ralph Nader: A Biography
Patricia Cronin Marcello

Carl Sagan: A Biography
Ray Spangenburg and Kit Moser

Sylvia Plath: A Biography
Connie Ann Kirk

Jesse Jackson: A Biography
Roger Bruns

Franklin D. Roosevelt: A Biography
Jeffery W. Coker

ALBERT EINSTEIN

A Biography

Alice Calaprice and Trevor Lipscombe

GREENWOOD BIOGRAPHIES

GREENWOOD PRESS
WESTPORT, CONNECTICUT . LONDON

Library of Congress Cataloging-in-Publication Data

Calaprice, Alice.
Albert Einstein : a biography / Alice Calaprice and Trevor Lipscombe.
p. cm. — (Greenwood biographies, ISSN 1540-4900)
Includes bibliographical references and index.
ISBN: 0-313-33080-8 (alk. paper)
1. Einstein, Albert, 1879–1955 — Biography. 2. Physicists—Biography.
I. Title. II. Series.
QC16.E5 C34 2005
530′.092—B 22 2004028175

British Library Cataloguing in Publication Data is available.

Library of Congress Catalog Card Number: 2004028175
ISBN: 0–313–33080–8
ISSN: 1540–4900

First published in 2005

Greenwood Press, 88 Post Road West, Westport, CT 06881
An imprint of Greenwood Publishing Group, Inc.
www.greenwood.com

Printed in the United States of America

The paper used in this book complies with the Permanent Paper Standard issued by the National Information Standards Organization (Z39.48-1984).

10 9 8 7 6 5 4 3 2 1

Although I am a typical loner in my daily life, my awareness of belonging to the invisible community of those who strive for truth, beauty, and justice has prevented me from feelings of isolation.

—From "My Credo/What I Believe," 1930

I firmly believe that love [of a subject or hobby] is a better teacher than a sense of duty—at least for me.

—To biographer Philipp Frank, 1940

Why is it that nobody understands me, yet everyone likes me?

—From an interview, *New York Times*, March 12, 1944

I have no special talents. I am only passionately curious.

—To biographer Carl Seelig, March 11, 1952

CONTENTS

SERIES FOREWORD

In response to high school and public library needs, Greenwood developed this distinguished series of full-length biographies specifically for student use. Prepared by field experts and professionals, these engaging biographies are tailored for high school students who need challenging yet accessible biographies. Ideal for secondary school assignments, the length, format, and subject areas are designed to meet educators' requirements and students' interests.

Greenwood offers an extensive selection of biographies spanning all curriculum related subject areas including social studies, the sciences, literature and the arts, history and politics, as well as popular culture, covering public figures and famous personalities from all time periods and backgrounds, both historic and contemporary, who have made an impact on American and/or world culture. Greenwood biographies were chosen based on comprehensive feedback from librarians and educators. Consideration was given to both curriculum relevance and inherent interest. The result is an intriguing mix of the well known and the unexpected, the saints and sinners from long-ago history and contemporary pop culture. Readers will find a wide array of subject choices from fascinating crime figures like Al Capone to inspiring pioneers like Margaret Mead, from the greatest minds of our time like Stephen Hawking to the most amazing success stories of our day like J. K. Rowling.

While the emphasis is on fact, not glorification, the books are meant to be fun to read. Each volume provides in-depth information about the subject's life from birth through childhood, the teen years, and adulthood. A thorough account relates family background and education, traces personal and professional influences, and explores struggles, accomplishments, and contributions. A timeline highlights the most significant life events against a historical perspective. Bibliographies supplement the reference value of each volume.

PROLOGUE: WHY EINSTEIN?

Pop idols, like fads, are here today and gone tomorrow. They are intensely popular for a relatively short time, have a short-term impact or influence on us—not always a positive one—and soon enough are forgotten and replaced by someone else. Once or twice in a century, perhaps, someone comes along in a special area of expertise who also has enduring qualities: such a person challenges our accepted ways of thinking, makes enormous and far-reaching contributions, and helps to revolutionize our world in a positive way. Almost everyone in the literate world recognizes the name and face of such a person. In modern times, one of these people was Albert Einstein, a binary star of superscientist and humanitarian and *Time* magazine's "Person of the Twentieth Century." This honor went to Einstein not to glorify him but to remind us of what he represented as a symbol of both human capabilities and frailties. Until Einstein, no one had challenged the accepted wisdom about the physical world so profoundly since Isaac Newton. Among other things, Einstein gave us new insights into the properties of space and time, showed that nothing can travel faster than the speed of light, predicted that time travel into the future is both mathematically and physically possible, and passionately, if perhaps sometimes naively, worked for world peace. For many, he is the standard of greatness.

Why does Einstein hold such fascination for us? What has made him such a captivating person for one hundred years, since the time he advanced his special theory of relativity in 1905 during his *annus mirabilis*, his "Year of Miracles"? Surely, his genius has something to do with his appeal; after all, he saw relationships in the physical world to which other people had been blind. But what made Einstein remarkable was not only his famous brain and his discoveries but many other characteristics as well: his charisma, humanity, modesty, wry sense of humor, courage to speak his mind even when his life was in danger; his love of children, music, and animals; and his resilience during hard times. We also know him for his well-known idiosyncrasies: the twinkle in his eyes, the shock of unruly hair, his dislike of socks, and his love of ice cream cones. Even though Einstein was not a perfect or consistent person, especially as a husband and father, and his idea of humor, often tongue in cheek, sometimes appears mean spirited, his frailties do not detract from his overall contributions to the world, and they punctuate his humanness. All these characteristics have made Einstein, despite his faults, a charismatic human being of tremendous intellect, wisdom, and depth.

To be a good scientist or mathematician, one does not have to be a genius. As Einstein said, one need only be very curious about the world. He was passionately devoted to his science, often to the exclusion of matters that are important in most people's lives. This behavior is not so strange, however: many talented people are obsessed with their gift, whether it is in science or sports—or even with a hobby. Einstein was more intensely preoccupied with his work in the first half of his life than later on, when he was past his prime in doing original work in physics. During his later years, he spent more time promoting the political, social, and educational causes he championed.

In this book, we present Einstein from birth to death and reflect on all the wonderful and terrible things that happened to him in between while he lived his life on two continents. Chapters 4, 7, and 12 explain his physics, and the remaining chapters cover the other important events in his life during the turbulent but interesting times through which he lived. Einstein will emerge not only as a great scientist but also as a humanitarian who was passionately concerned about the welfare of his fellow human beings and the security of the world. He respected people of all occupations, of all races, and of all ages. His advice was always to do the best one can no matter what one chooses to do in life; above all, he said, we have an obligation to do what is best for our community.

Albert Einstein could have rested on his laurels after relativity theory and, later, the Nobel Prize made him internationally famous, but he chose not to. He had more to do.

* * *

We would like to thank the following people for their input and support: AC is grateful to a British friend who read the first draft and tried telling her how they write and punctuate English over there; TL tried telling her again, but her many years of consorting with the *Chicago Manual of Style* and Follett's *Modern American Usage* left her unconvinced. Both of us thank Kevin Downing, our editor and old friend at Greenwood, for presenting us with the challenge of writing this book; and we thank the remaining colleagues, friends, and family members who egged us on by never failing to ask if we'd finished the book yet.

TIMELINE

1879 March 14	Albert Einstein is born in Ulm, Germany, to Jewish parents, Hermann and Pauline Koch Einstein.
1880	The Einstein family moves to Munich.
1881	Einstein's sister, Maja, is born.
1884	Young Einstein is shown a compass by his father, which fascinates him, making him aware of forces that can't be seen.
1885	In the fall, Einstein begins his education at a Catholic neighborhood school, the only Jewish child in his class. He receives Jewish religious instruction at home and becomes curious about religion; he also begins violin lessons.
1888	As Einstein becomes nine years old, he enters the Luitpold-Gymnasium (secondary school) in Munich.
1889	At the age of 10, Einstein's interest in physics, mathematics, and philosophy begins when a young family friend who recognized Albert's intelligence and curiosity introduces him to these subjects through a number of popular scientific and technical books. In April, a baby named Adolf Hitler is quietly born in Austria.
1890	By this time, Albert is able to prove the Pythagorean theorem by himself, and he enjoys working out difficult problems and puzzles.
1891	Reading on his own, Albert teaches himself higher

	mathematics and calculus.
1892	Young Albert is becoming a good violinist and continues to read science books voraciously.
1894	Einstein's family moves to Italy; he stays in Munich to finish school but is unhappy and quits, joining his family at the end of the year. His teacher had told him that he would never amount to anything and that because of his irreverence, his presence undermined the whole class's respect for the teacher.
1895	Albert tries to enroll in the Swiss Federal Polytechnical School in Zurich two years early, but he fails the nonscientific part of the entrance exam and is urged to spend another year in secondary school.
1896	At the age of 17, Einstein relinquishes his German citizenship, with his father's consent, because he detests the country's obsession with regimentation in most aspects of life, and he remains stateless for the next five years. He enters the Swiss Federal Polytechnical Institute in October. He begins a relationship with Mileva Marić, a physics student from Serbia.
1899	Einstein applies for Swiss citizenship and spends his summer vacation with his mother and sister in Switzerland.
1900	Einstein is graduated from the Federal Polytechnical Institute and begins a job search in Europe. At the same time, he works on problems in theoretical physics that are of interest to him.
1901	Becomes a Swiss citizen. Seeks employment. His first scientific paper, "Conclusions Drawn from the Phenomena of Capillarity," is published in March in *Annalen der Physik.* In the summer, works as a substitute teacher at the technical school in Winterthur and in the fall as a tutor in a private boarding school in Schaffhausen. Stays in touch with and visits Mileva regularly. Begins work on a doctoral dissertation on molecular forces in gases that he submits to the University of Zurich in November. December, applies for a position at the Swiss Patent Office in Bern.
1902	Probably in January, daughter Lieserl is born out of wedlock to Mileva. Einstein withdraws his doctoral

dissertation from the University of Zurich. June, begins a provisional appointment as Technical Expert, Third Class, at the Patent Office in Bern. October, his father dies in Milan.

1903 January 6, marries Mileva in Bern, where they take up residence. September, daughter Lieserl is registered, which may have indicated intention to put her up for adoption in case knowledge of the illegitimacy would be a threat to Einstein's federal appointment. No mention is made of Lieserl after she contracts scarlet fever in September while Mileva is on a visit to Budapest. (Lieserl never lived with her parents, Einstein never saw his daughter, and all trace of her has been lost.) At this time, Mileva is pregnant again.

1904 May 14, son Hans Albert ("Adu") is born in Bern (died 1973 in Falmouth, Massachusetts; buried in Woods Hole, Massachusetts). September, Einstein's provisional appointment at the Patent Office becomes permanent.

1905 Einstein's "year of miracles" with respect to his scientific publications. April 30, submits his doctoral dissertation, "A New Determination of Molecular Dimensions," for publication. In addition, publishes three of his most important scientific papers: "On a Heuristic Point of View concerning the Production and Transformation of Light" (published June 9), which deals with the quantum hypothesis, showing that electromagnetic radiation interacts with matter as if the radiation has a granular structure (the so-called photoelectric effect); "On the Movement of Small Particles Suspended in Stationary Liquids Required by the Molecular-Kinetic Theory of Heat" (published July 18), his first paper on Brownian motion, leading to experiments validating the kinetic-molecular theory of heat; and "On the Electrodynamics of Moving Bodies" (published September 26), his first paper on the special theory of relativity and a landmark in the development of modern physics. A second, shorter paper on the special theory, published November 21, contains the relation $E = mc^2$

in its original form.

1906 January 15, formally receives doctorate from the University of Zurich. March 10, promoted to Technical Expert, Second Class, at the Patent Office.

1907 While still at the Patent Office, seeks other employment, including at the cantonal school in Zurich and at the University of Bern.

1908 February, becomes a *Privatdozent* (lecturer) at the University of Bern. Sister Maja receives her doctorate in Romance languages from the University of Bern.

1909 May 7, is appointed Extraordinary Professor of Theoretical Physics at the University of Zurich, effective October 15. Resigns from his positions at the Swiss Patent Office and the University of Bern. Receives his first honorary doctorate, at age 30, from the University of Geneva.

1910 March, sister Maja marries Paul Winteler, son of Einstein's landlord in Aargau. July 28, second son, Eduard ("Tete"), is born (died 1965 in a psychiatric hospital in Burghölzli, Switzerland; he had had a history of schizophrenia since he was in his twenties). October, completes a paper on critical opalescence and the blue color of the sky, his last major work in classical statistical physics.

1911 Accepts an appointment as director of the Institute of Theoretical Physics at the German University of Prague, effective April 1, and resigns his position at the University of Zurich. Moves his family to Prague. October 29, attends the first Solvay Congress in Brussels.

1912 Becomes reacquainted with his divorced cousin Elsa Löwenthal and begins a romantic correspondence with her as his own marriage disintegrates. Accepts appointment as professor of theoretical physics at the Eidgenössische Technische Hochschule (ETH) in Zurich (formerly the Polytechnical Institute), beginning in October, and resigns his position in Prague.

1913 September, sons Hans Albert and Eduard are baptized as Orthodox Christians near Novi Sad, Hungary (later Yugoslavia), their mother's hometown.

	November, is elected to the Prussian Academy of Sciences and is offered a position in Berlin, home of Elsa Löwenthal. The offer includes a research professorship at the University of Berlin, without teaching obligations, and the directorate of the soon-to-be-established Kaiser Wilhelm Institute of Physics. Resigns from the ETH.
1914	April, arrives in Berlin to assume his new position. Mileva and the children join him but soon return to Zurich because of Mileva's unhappiness in Berlin. August, World War I begins.
1915	Cosigns a "Manifesto to Europeans" upholding European culture, probably his first public political statement. November, completes his work on the logical structure of general relativity.
1916	Publishes "The Origins of the General Theory of Relativity" (later to become his first book) in *Annalen der Physik*. May, becomes president of the German Physical Society. Publishes three papers on quantum theory.
1917	February, writes his first paper on cosmology. Becomes ill and is weakened by a liver ailment and an ulcer. Elsa takes care of him. October 1, begins directorship of the Kaiser Wilhelm Institute of Physics. After World War I, holds dual Swiss and German citizenship.
1919	February 14, is divorced from Mileva. Divorce decree stipulates that the interest from any future Nobel Prize monies go to her and the children for living expenses and to ensure their permanent financial security. May 29, during a solar eclipse, Sir Arthur Eddington experimentally measures the bending of light and confirms Einstein's predictions; Einstein's fame as a public figure begins. June 2, marries Elsa, who has two unmarried daughters, Ilse (22 years old) and Margot (20 years old), living at home. Late in the year, becomes interested in Zionism through his friendship with Kurt Blumenfeld.
1920	February 20, mother dies in Berlin. Expressions of anti-Semitism and anti–relativity theory become

	noticeable among Germans, yet Einstein remains loyal to Germany. Becomes increasingly involved in nonscientific interests.
1921	April and May, makes first trip to the United States. Receives honorary degree and delivers four lectures on relativity theory at Princeton University as part of the Stafford Little Lectures, which Princeton University Press in the United States and Methuen and Company in Great Britain later publish as *The Meaning of Relativity*. Accompanies Chaim Weizmann on U.S. fund-raising tour on behalf of Hebrew University of Jerusalem.
1922	Completes his first paper on a unified field theory. October through December, takes trip to Japan, with other stops en route to the Far East. November, probably while en route to Shanghai, learns that he has won the 1921 Nobel Prize in physics.
1923	Visits Palestine and Spain.
1924	Stepdaughter Ilse marries Rudolf Kayser, a journalist and future Einstein biographer. Ilse had, for a time, considered marrying Einstein, who appears to have been in love with her, before he married her mother.
1925	Travels to South America. In solidarity with Gandhi, signs a manifesto against compulsory military service. Becomes an ardent pacifist. Receives Copley Medal. Until 1928, serves on board of governors of Hebrew University.
1926	Royal Astronomical Society of England awards him its gold medal.
1927	Son Hans Albert marries Frieda Knecht over his father's objections.
1928	Becomes ill again, this time with a heart problem. Is confined to bed for several months and remains weak for a year. April, Helen Dukas is hired as his secretary and remains with him as secretary and housekeeper for the rest of his life.
1929	Begins lifelong friendship with Queen Elisabeth of Belgium. June, receives Planck Medal.
1930	First grandchild, Bernhard, is born to Hans Albert and Frieda. Stepdaughter Margot marries Dmitri

	Marianoff (marriage later ends in divorce). Signs manifesto for world disarmament. December, visits New York and Cuba and stays (until March 1931) at the California Institute of Technology (Caltech) in Pasadena.
1931	Visits Oxford in May to deliver the Rhodes Lectures and receives honorary degree, then spends several months at his summer cottage in Caputh, southwest of Berlin. December, en route to Pasadena again.
1932	January through March, visits Caltech again. Returns to Berlin. Later, agrees to accept an appointment as professor at the Institute for Advanced Study in Princeton, New Jersey, at this point in the planning stages with no campus. December, makes another visit to the United States.
1933	January, Nazis come to power in Germany. Resigns membership in the Prussian Academy of Sciences, gives up German citizenship (remains a Swiss citizen), and does not return to Germany. Instead, from the United States, goes to Belgium with Elsa and sets up temporary residence at Coq sur Mer. Ilse, Margot, Helen Dukas, and Walther Mayer, an assistant, join them, and security guards are assigned to protect them. Takes trips to Oxford, where he delivers the Herbert Spencer Lecture in June, and Switzerland, where he makes his final visit to son Eduard. Rudolf Kayser, Ilse's husband, manages to have Einstein's papers in Berlin sent to France and eventually brought to the United States. Early October, leaves Europe, together with Elsa, Helen Dukas, and Walther Mayer, and arrives in New York on October 17 on the *Westmoreland;* Ilse and Margot and their husbands remain in Europe. Publishes, with Sigmund Freud, *Why War?* Begins professorship at the Institute for Advanced Study, temporarily located in the old Fine Hall (now Jones Hall) on the Princeton University campus.
1934	July 10, Ilse dies in Paris at age 37 after a long and painful illness. Margot and Dmitri come to Princeton. Rudolf remains in Europe.
1935	Fall, moves to 112 Mercer Street, Princeton, where

	Einstein, Elsa, Margot, Maja, and Helen Dukas will live out their lives. Receives Franklin Medal.
1936	Hans Albert receives doctorate in technical sciences from the ETH in Zurich (in 1947 he becomes a professor of hydraulic engineering at the University of California at Berkeley). December 20, Elsa dies after a long battle with heart and kidney disease.
1939	Sister, Maja Winteler-Einstein, comes to live at Mercer Street. August 2, signs famous letter to President Roosevelt on the military implications of atomic energy. World War II begins in Europe.
1940	Receives U.S. citizenship. Maintains dual U.S. and Swiss citizenship until his death. Citizenship had been proposed earlier by an act of Congress, but Einstein preferred waiting to be naturalized the customary way.
1941	December, United States enters World War II.
1943	Becomes consultant to U.S. Navy Bureau of Ordnance, Section on Explosives and Ammunition.
1944	A newly handwritten copy of the original 1905 paper on the special theory of relativity is auctioned for $6 million as a contribution to the war effort.
1945	World War II ends. Retires officially from the faculty of the Institute for Advanced Study, receives a pension, but continues to keep an office there until his death.
1946	Maja suffers a stroke and is confined to bed. Einstein becomes chairman of the Emergency Committee of Atomic Scientists. Urges United Nations to form a world government, declaring that it is the only way to maintain world peace.
1948	August 4, Mileva dies in Zurich. December, Einstein's doctors tell him that he has a large aneurysm (abnormal dilatation) of the abdominal aorta.
1950	March 18, signs his last will, naming his friend Otto Nathan as executor and Otto Nathan and Helen Dukas as trustees of his estate. His literary estate (the archive) is to be transferred to the Hebrew University of Jerusalem after the death of Nathan and Dukas. (Arrangements are later made for an earlier transfer.)

1951	June, Maja dies in Princeton.
1952	Is offered the presidency of Israel, which he declines.
1954	Develops hemolytic anemia.
1955	April 11, writes last signed letter, to Bertrand Russell, agreeing to sign a joint manifesto urging all nations to renounce nuclear weapons. April 13, aneurysm ruptures. April 15, enters Princeton Hospital. April 18, Albert Einstein dies at 1:15 A.M. of a ruptured arteriosclerotic aneurysm of the abdominal aorta, caused by hardening of the arteries. He had opposed surgery to prolong his life.

Chapter 1

"IT IS A KNOWN FACT THAT I WAS BORN..."

My life is a simple thing that would interest no one. It is a known fact that I was born, and that is all that is necessary.

—To a Princeton (New Jersey) High School reporter, in the school's newspaper, *The Tower*, April 13, 1935

There was much commotion in the Einstein household in the southern German town of Ulm that Friday in 1879. It was March 14, and a baby boy had been born in the house on Bahnhofstrasse earlier that day. The parents, Pauline and Hermann, were elated. The jovial and easygoing Hermann, wearing his pince-nez eyeglasses and sporting a mustache, looked down adoringly at his young wife and the newborn resting in bed, and, like all fathers, felt full of hope for a good life for the little boy, his first child. The parents' only concern was that the baby's head was unusually large and angular, and Mrs. Einstein was afraid it signaled a birth defect. The doctor assured her that all would be well, and, sure enough, in time the shape of his head looked normal. Pauline and Hermann named their new baby "Albert."

When the Einsteins were first married in 1876, they had lived in the even smaller town of Buchau, 30 miles southwest of Ulm, in the state of Württemberg in southern Germany. Hermann's family had lived there since at least the 1750s, while Pauline's came from a town some 50 miles farther northwest of Ulm in Cannstatt near the large city of Stuttgart. On the day of their wedding, Pauline was 18 and Hermann was 29. Although both were of Jewish origin, they were not particularly attached to Jewish religious traditions. Instead of observing the Jewish custom of naming a child after a beloved relative, they used only the first letter of the grandfather's name, Abraham, perhaps as a statement of their assimilation into the broader society. But they retained the deep respect for education and humanitarianism that had been common among Jewish families worldwide for centuries.

Hermann, a happy-go-lucky man who had already failed in the featherbed business, had set up a small electrical and engineering workshop on the south side of the cathedral two years earlier. His family was industrious but not wealthy, so Pauline's helped him get started by providing financial support. He was aspiring to make a decent living for his family at a time when Europeans and Americans were pioneering the wider use of electricity as lighting. Although the existence of electricity had been known for over two and a half millennia—Thales of Miletus discovered that amber could attract light objects when it is rubbed, and the ancient Greek word for amber is "elektron"—not much practical application had yet been found for it beyond the telegraph and telephone. The year of Albert's birth, though, Thomas Alva Edison, "the wizard of Menlo Park," was working on a patent for the electric lightbulb. The Siemens company of Germany was about to exhibit the first electric tram in Berlin that year, and the promise of further electrical applications lay ahead.

Unfortunately, Hermann's sense for business was not as good as his sense of humor, and his optimistic disposition invigorated neither research nor sales. The small enterprise collapsed before Albert even reached his first birthday. Desperate but not discouraged, Hermann decided that the family had to move to a larger town if he wanted to become successful in achieving his goals. After talking things over with his brother Jacob, an engineer, they decided in June 1880 to move southeast to Munich, the capital of the state of Bavaria. Here, in this large and sophisticated city of overwhelmingly Catholic residents, the two brothers opened a small electrochemical works around the time the little Einstein son became one year old. Albert, therefore, retained no memories of Ulm, his birthplace.

Until a few years before Einstein's birth, Württemberg and Bavaria had been independent kingdoms. Then, in 1871, along with other independent German states, they were absorbed into the kingdom of Prussia and became part of a new German empire. The leader of the new nation, Otto von Bismarck, was from Prussia, whose residents were known for their strict discipline, their unquestioning obedience to their elders and to authority, and their deep respect for those of high standing in society. The new regime soon forced the easygoing southerners into an unfamiliar way of life that also crept into the city school systems, a change that would profoundly affect the young Albert.

In Munich, meanwhile, Pauline set up her household in a small rented house, waiting to see what fate had in store for the family business. As it turned out, Hermann seemed to have learned from his past mistakes, and the business took off and flourished. Within five years, the family was able to move into a larger home on the outskirts of the city. In the meantime, however, they were happy in the small quarters, especially when a new member of the family arrived about a year and a half later, in November 1881: a baby sister for Albert named Maja. Albert's parents told him that he would now have something new to play with, but when Albert, then two-and-a-half, was shown the new baby, he commented in disappointment, "Yes, but where are its wheels?"

On the surface, Albert's development in his early years did not foreshadow a genius in the making. He was slow to talk, not beginning to do so until around the time that Maja was born, but then he seemed to make up for lost time by speaking in complete sentences. For several years, however, he would first say his words quietly to himself before saying them out loud. Therefore, to some people it sounded as if he said everything twice, once softly and once loudly. Some medical practitioners, noting that Albert did not become fluent in speech until around the age of 10, now think that he may have had a form of childhood dyslexia. Even later in life, he admitted that he had always had a poor memory for words. In fact, he told a psychologist during an interview for a book, "I very rarely think in words at all. A thought comes, and I may try to express it in words afterwards." Perhaps that's why he later became famous for his creative "thought experiments."

Little Albert often played alone, content to amuse himself with building blocks and puzzles. But he also had a strong temper. When he became angry, his whole face turned yellow except for the tip of his nose, which turned white, and he would lose control of himself. Once, during a tantrum, he vented his anger at his violin teacher by throwing a chair at her, and at other times he assaulted little Maja with a trowel and threw a skittles ball *(Kegelkugel)* at her head. (In the early days of Christianity, German monks had carried a *Kegel*, or club, for self-defense, and in their games, the *Kegel* represented a sin or temptation, and the monks would throw stones at it until they knocked it over. Today's Germans play *Kegelen* with nine small bowling pins and a ball.)

In her short biography of her brother, Maja would later recall that, on the other hand, Albert would also spend hours patiently building houses of cards up to 14 stories high, undeterred by any upsets he may have encountered during his play. He was always curious about new things his parents or others brought for him and persistent and patient in learning to use them. When he was older, he would claim that he was not smarter than others, only more passionately curious; he also said that his natural childhood curiosity lasted far into his adulthood. He wrote to a colleague, "I ... developed so slowly that I did not begin to wonder about space and time until I was an adult. I then delved more deeply into the problem than any other adult would have done." He thereby expressed his belief that it is usually children, not adults, who reflect on the kinds of problems that became his life's work.

While stressing that it is impossible to make a proper diagnosis, a professor at Cambridge University in England who researches and treats autism has recently suggested that some of Einstein's unusual behavior as a child and later in life was symptomatic of Asperger's disease, a mild and "high-functioning" form of autism. Throughout history, myths have grown up around the great, seeking to account for what made them so wise or so brave in battle. Some of these guesses may be right, while others are most certainly wrong. Einstein researchers consider that in Einstein's case, the possibility of a mental disorder is highly unlikely. There are a number of Einstein myths that may fit an Asperger's diagnosis: that he was socially inept and awkward, had difficulty writing, hardly spoke, and was a notoriously confusing lecturer. None of this is true, even though, as with all of us, these traits may have been evident from time to time throughout his long life. Moreover, a psychiatrist at the University of California, San Diego, responding to the Cambridge professor, commented that a good sense of humor was a trait not seen in people with severe Asperger's. Einstein was noted for his wit. It could be charming, too. Later in life he noticed in Princeton that his cat, Tiger, seemed depressed on a rainy day and commiserated, "I know what's wrong, dear fellow, but I don't know how to turn it off."

One day, when Albert was about five years old, his father pulled out from his pocket a small instrument that would have a profound effect on his son's life: a compass. The consistent northward swing of the needle struck Albert with awe and wonder as he realized even at such a young age that there are forces in nature that one cannot see. This magic left a lasting impression on him, one that he would recall often in his later years. That same year, while he was still under the age of admission to a Munich primary school, Albert began his education at home with a tutor.

At the age of six, in the fall of 1885, Albert entered the Catholic primary school in his neighborhood, probably beginning in the second grade. He was the only Jewish child in class. Religious instruction was part of the school curriculum, so he became acquainted with the Bible stories and the saints. At home he was given lessons in the Jewish tradition by a distant relative. His parents, who were indifferent to religion, did not feel competent enough to do so, yet they wanted to give their children the chance to learn about their heritage. During this time, Albert became quite interested in conventional religion, wanting above all to please God. Much later in life, while supplying "autobiographical notes" for the volume that bears his name in the Library of Living Philosophers (1949), he wrote, "Thus I came ... to a deep religiosity, which, however, reached an abrupt end at the age of twelve." By using the word "religiosity," he was distancing himself from what he considered popular superstition. To the end of his days, he regarded himself as "religious" in a much broader sense and took pains to explain what he meant by that. (See more about this at the end of chapter 8.) On the matter of a personal God, he declared himself agnostic.

Albert also began to receive lessons in playing the violin around the age of six, lasting until he was 14. Music had an especially calming effect on the sometimes temperamental child. He proved to have musical talent and enjoyed music until the end of his life both as a listener and as a player of the violin and later the piano. "If I were not a physicist," he would say in an interview in 1929, "I would probably be a musician. I live my daydreams in music. I see my life in terms of music. I get most of my joy in life out of music." Some people have reported that Einstein was quite a good musician, but others weren't so enthusiastic. A professional violinist claimed he "fiddled like a lumberjack"; a famous pianist playing with him demanded, "For heaven's sake, Albert, can't you count?"; and a music critic in Berlin, thinking Einstein was famous for his violin playing rather than physics, judged that "Einstein's playing is excellent, but he does not deserve his world fame; there are many others just as good." As to Einstein's alternative career choices, he would later also say he would rather choose to become a plumber or a salesman, and he even suggested to science students that being a lighthouse keeper had its advantages—obviously he was a man of many interests. He chose or recommended these professions because he believed some serious thinking could be done while holding these jobs.

When he was nine, in the fall of 1888, Albert entered the first year of the nine-year program of the German secondary school, or Gymnasium, in Munich, which put him on a track to enter a university when he was 18. Here he met some Jewish children, but he did not form any particularly close friendships, remaining somewhat aloof and a loner, not endearing himself to either the children or his teachers. Indeed, one of his teachers told him he would never amount to anything and that his mere presence undermined the class's respect for the teacher. Another accused him of having a memory like a sieve.

In those days in Germany, teachers were rigid and authoritarian and demanded the utmost respect even if some of them may not have earned it. The German school system enforced strict rules of behavior and followed a specific and tightly controlled curriculum that largely depended on memorization. It was a common practice to humiliate unprepared students in front of the class and to apply corporal punishment as was deemed necessary. "To me, the worst thing is for a school principally to work with the methods of fear, force, and artificial authority. Such treatment destroys the sound sentiments, the sincerity, and the self-confidence of the pupil," Einstein would write later. "Humiliation and mental oppression by ignorant and selfish teachers wreak havoc in the youthful mind that can never be undone and often exert a baleful influence in later life." Furthermore, he found the militaristic discipline distasteful, as he would also find the military life itself as he became a pacifist: "That a man can take pleasure in marching in formation to the strains of a band is enough to make me despise him." The strict, narrow-minded approach to education was incompatible with Albert's independent-thinking personality. A boy like him who would not have aspired to be a teacher's pet is likely to have elicited undeserved negative comments from those in authority.

Still, he did have a favorite teacher, his homeroom teacher who also taught history, Latin, and Greek during his fourth and sixth years in the school. Albert didn't even mind being kept after school by him and considered it a pleasure. Furthermore, his Gymnasium was one of the better and larger ones in Germany, and at the time he attended it, it had a reputation as an enlightened school with a liberal atmosphere. Maybe Albert's displeasure and bad memories had more to do with his own personality or personal experiences than with his school. He may also have provoked the teachers who scolded him, or his memory may have been colored by the generally oppressive mood in Germany.

From this time on, until he left the Gymnasium, his interest in physics, mathematics, and philosophy developed independently of his formal studies in school. His uncle Jacob, an engineer, and Max Talmey, a young medical student who had dinner at the Einsteins' house once a week beginning in the fall of 1889 until 1894, were great influences on Albert during his formative years. They encouraged his inherent and insatiable curiosity about everything from the little compass to religion. Talmey brought along popular science books that he thought would interest the curious young boy. He also brought stacks of philosophy books, and the two of them discussed many of the questions raised by Albert, who was thirsting for knowledge about the world. To Albert's delight, Talmey treated him as an equal despite their difference in age. His investment of time in Albert came at a crucial age, at a time when a young boy matures into a thinking adult. At this point, Albert became interested not only in mathematics but in all the natural sciences as well.

By the time Albert was 12, he had read so much science and mathematics that he now questioned the often miraculous accounts of events and people he read in the Bible. Lacking the faith to believe them, he ended his acceptance of traditional religion. He felt there was a conspiracy by the state to lie to young people about the world as it really is, and he would have none of it, though later in life he would develop some personal religious beliefs. For now, he continued to study science and mathematics—subjects he found beyond reproach—on his own. He especially prized his "sacred little geometry book," a small book that dealt with plane geometry that was probably given to him by Uncle Jacob. The clear expositions in the book had a great impact on Albert, causing him to experience his "second wonder," after the compass his father had shown him many years earlier. Later he received more advanced books and became proficient in analytical geometry and calculus. Soon he was able to prove mathematical theorems and solve difficult problems, even finding an entirely original proof for the Pythagorean theorem in geometry. He found his greatest happiness in solving these problems and had little interest in pursuing close friendships with his schoolmates. Like Isaac Newton before him, he was attracted by the synthesis of classical geometry because it was so clear-cut and certain. Remembering his fondness for mathematics as a child, he later, in 1935, told a reporter for a high school newspaper in Princeton, New Jersey, "As a boy of twelve years making my acquaintance with elementary mathematics, I was thrilled in seeing that it was possible to find out truth by reasoning alone. … I became more and more convinced that even nature could be understood as a relatively simple mathematical structure."

Rumor has had it that young Albert was not a good student, but this is a myth. Some of this fiction may be due to the negative comments his teachers made, but more likely it was the change in the grading system at his school. On a scale of 1 to 6, at one time "1" was considered the best, but later the system was reversed. He always had outstanding grades in mathematics and physics and above-average grades in his other school subjects. His one weakness, which he admitted freely, was foreign languages; in fact, he never mastered the English language after he came to America, no doubt because he was already middle aged by then. As a young student, he showed no sign of real "genius." His gifts did not become apparent until he did his seminal work in physics in his mid-twenties and later, when he looked at time and space in a completely new way.

By the age of 13, young Einstein became more seriously interested in both philosophy and music. He studied German philosopher Immanuel Kant's *Critique of Practical Reason*, whose moral theory of freedom of the will appeared to resonate with him. There he probably read one of Kant's most famous declarations, written in the conclusion of the book: "Two things fill the mind with ever new and increasing wonder and awe: the starry heavens above and the moral law within." Indeed, these words are often attributed to Einstein himself. At this time, too, the teenager gave up taking violin lessons, which he found boring and too mechanical. He began to study and practice music on his own, however, thereby gaining deeper satisfaction from his playing and listening. His favorite music then and throughout his life were the Mozart sonatas, which he considered so pure and beautiful that he saw them as a reflection of the inner beauty of the universe. When he had a son of his own, he told him, "Stick to the Mozart sonatas. Through them, your papa also learned to know music well."

In 1894, Albert's family decided, with a heavy heart, to leave Munich. Hermann and Jacob's company had been prospering by installing streetlights, illuminating various projects around the cities, and manufacturing lamps, electric meters, dynamos, and other equipment necessary for urban electrification. At its peak, the company employed almost 200 people, but soon the large electrical engineering firms moved in on the business, outbidding the Einsteins for the municipal contracts that had been their mainstay. They decided to abandon their enterprise in Germany and try their luck in northern Italy, in Pavia near Milan. Pulling up their roots was painful for all of them, especially for Albert and Maja, who sadly watched their home being dismantled. Albert, a sensitive and perceptive boy, was 15 at the time and aware of the loss of the important business contract that would have kept his family afloat in Germany. Sensing latent anti-Semitism in the transaction, he began to adopt a negative attitude toward his homeland.

Because Albert's parents did not want to interrupt his studies at the Gymnasium, where he would soon begin his seventh year, they arranged for him to stay with a distant relative in the city, an elderly woman, but took the younger Maja with them. Albert was miserable in his new situation. It was enough that he hated his school, but now he no longer had the comfort of his congenial family life and home, for which he yearned. After several months of anguish and reading his parents' letters of the happy life they enjoyed in Italy, he quit school without his parents' permission. All it took was a sick note from the family doctor saying he suffered from "nervous exhaustion." At the end of December, he abruptly boarded a train to Milan, where his parents were living temporarily before their move to Pavia.

Pauline and Hermann were no doubt surprised and conflicted when they found Albert on their doorstep, begging to stay with them in Milan. He promised to study on his own to prepare himself for the fall 1895 entrance examination to the Swiss Federal Institute of Technology in Zurich, which he wanted to attend, perhaps at the suggestion of Uncle Jacob or his mother, a down-to-earth woman who hoped her son would choose a practical profession. This institution was one of the best in Europe for those who were interested in science and technology, and it did not require a Gymnasium diploma if a student received good marks on an entrance exam. His indulgent and trusting parents took him at his word.

Albert loved living in Italy, and he studied diligently all year, also helping his uncle with design work for his equipment. The boy's study habits were quite strange, though. Even when he was in a noisy situation, he would be able to withdraw from the group and sit on a sofa, "take pen and paper in hand, set the inkstand precariously on the arm rest, and lose himself so completely in a problem that the polyphonic conversation stimulated rather than disturbed him," according to his sister's recollections. Albert was serious about his goals, and now, for his own pleasure, he wrote his first scientific essay, reflecting his interest in electromagnetic phenomena and showing that he mastered much of the current knowledge about electrodynamics.

At this time, when Albert was only 16 years old, he first envisioned an experiment that would later lead to his special theory of relativity. He imagined what it might be like to chase a beam of light: could he go as fast as the beam, or would he be able to go faster and overtake it? "If I pursue a beam of light with the velocity c (the velocity of light in a vacuum), I should observe such a beam of light as an electromagnetic field at rest, though spatially oscillating. ... From the very beginning it appeared to me intuitively clear that, judged from the standpoint of such an observer, everything would have to happen according to the same laws as for an observer who, relative to earth, was at rest," he would write toward the end of his life in his "Autobiographical Notes." What would the world look like to an observer riding on a beam of light, traveling at 186,324 miles per second? Would the light emitted from a moving train be faster than the light from a train at rest in a station? Later, with his special theory of relativity (see chapter 4), he became the first person to recognize that light always travels at the same speed, at least through a vacuum, no matter how fast the object from which the light is being measured may be moving. That is, the light does not travel at the speed of light plus the speed of the moving train: it remains constant. This discovery has had great implications for our view of time and space.

That summer, the Einstein family moved from Milan to Pavia, and Albert, when not studying, spent time hiking from there across the Alps to visit relatives in Genoa. (His parents would move back to Milan in the fall of the following year.)

In October, Albert went by train to Zurich to take the entrance exam for admission to the Poly. He was two years below the normal age for admission, having gained special permission through the efforts of a family friend to take the exam early. He passed the mathematical and science parts with high marks, which impressed one of the professors so much that he invited Albert to stay in Zurich and attend his classes. But the Poly's director, Albin Herzog, felt he was not yet quite ready to start his university career, even though he may indeed be a Wunderkind. He wanted Albert to wait at least another year to better prepare himself for the challenges that lay ahead and to fill the gaps in his studies, primarily in languages and history. Gustav Meier, a family friend, recommended that he go to Switzerland and attend the technical division of a high school there for the school year 1895–1896, and his parents agreed. The school year was about to start in Aarau, Switzerland, a village near Zurich in the canton of Aargau, and Albert quickly agreed to heed his elders' advice to attend. (A "canton" is a state in the Swiss confederation.) In a year, he could graduate and thereby satisfy the entrance requirements for the Poly without taking the exam again. His goal was to become more competent in nonscientific subjects and in chemistry and to allow himself another year to grow up before starting the serious business of university life.

Chapter 2

TRAINING THE MIND

> The value of an education … is not the learning of many facts, but the training of the mind to think about things that cannot be learned from textbooks.
>
> —Commenting on Thomas Edison's opinion that a college education is useless, 1921

This last year of school before his university education began turned out to be wonderful and instructive for Albert in every way. The Swiss school system suited him much better. Gone was the authoritarian, stuffy, and formal attitude and regimentation he had detested in Germany, and in its place he found an informal and friendly atmosphere conducive to learning without fear of strict reprisals. The administrators and teachers treated the students as individuals rather than a military-like unit, encouraging them to be independent in both their thought and their actions. The teachers, in turn, earned the students' respect as individuals, and the young people went to them freely and fearlessly with their questions and problems.

Switzerland was also a peaceful and neutral country that had no desire to go to war, conquer peoples, or gain new territory, even though its soldiers and army—which existed only for defensive purposes—were supposed to be among the best in the world. The small nation was therefore the ideal environment for someone of Albert's character, talents, and growing political interests, and he thrived in his new surroundings. Much of this newfound joy was made possible by Jost and Pauline Winteler, with whom he boarded while attending the cantonal school that year.

The cantonal school system of Aargau had a tradition of championing a liberal and secular education and was also known for its physics and engineering laboratory. The town of Aarau was the capital of the fertile, northern canton in the area where the river Aare is joined by the Suhr River and then flows northeast into the Rhine. Most of the 200,000 people of the canton spoke German, and they were engaged in tobacco growing, silk-ribbon weaving, and straw plaiting. The town of Aarau, about half an hour by train from Zurich, was small, with around 7,000 residents. The cantonal school, which served as a prep school for the Poly in Zurich, consisted of a section where the humanities were taught, a two-year commercial school, and a technical or trade school, which Albert attended along with 64 other students.

Jost Winteler, a popular teacher of Greek and history at the Gymnasium, was a warm and friendly man and the patriarch of a large family of seven children. Albert soon felt great affection for him and Pauline, referring to them as "Mama" and "Papa" Winteler. He also respected Jost's liberal political and religious views, later calling his distrust of imperial Germany prophetic. During this year, too, Albert took a liking to their pretty daughter, Marie, who was two years older than he. After the young teenager learned of Marie's love of the violin, he tried to impress her with his violin playing. She must have liked what she heard because she soon agreed to become his first girlfriend. When he returned home to Pavia to spend Easter vacation with his parents, the lonely Albert wrote to Marie, telling her how much he missed her and adding, "Love brings much happiness, much more so than pining for someone brings pain." Both sets of parents were pleased about the romantic affections of the young couple, even though it lasted for only a year, when Albert ended it with some guilt. After he left Aargau, he continued to stay in touch with the family. Tragedy befell the Wintelers, however, when in 1906 Marie's brother Julius shot and killed his mother, the husband of his sister Rosa, and then himself. One of Marie's other brothers, Paul, married Albert's sister Maja in 1910.

In early 1896, just three months after Einstein arrived in Aarau, the self-assured young man, with his complaisant father's consent, took a bold step: he decided to give up his German citizenship. Albert had been in Switzerland for only three months, and already he felt at home. He did not have fond memories of Germany, probably in part because of his unhappy experiences in his schools, his sensitivity to criticism, and his family's business failings. While he was at it, he also decided to renounce any religious identity. He felt that Judaism as he understood it offered him nothing that seemed to be of value, and from now on declared himself as having "no religious affiliation," though he renounced only his religion, not his forebears. His pacifist feelings were evolving, too, and he said he did not want to be part of a country where militarism, which had spread from Prussia to infect the entire nation, ranked supreme. He also feared that he might be drafted into the German army, for which he was totally unsuited. In sum, he had quickly become enamored of the democratic and civilized way of life in Switzerland, where he seemed to breathe a more liberal air that suited him, and he was ready to wait the required five years to become a citizen. With this renunciation of his fatherland, he became stateless and then remained without any kind of citizenship throughout his college years.

At Aarau, Albert was considered a little unusual by his fellow students because he didn't want to study one of the more popular subjects in college. Most of the students wanted to study law or medicine or become teachers, but he planned to devote himself to mathematics and physics. He began to think seriously about what lay ahead of him, even though he believed that "a happy man is too satisfied with the present to think too much about the future," as he said in an essay for his final exam in French class on "My Plans for the Future" while setting down his thoughts on his strengths and weaknesses. He wrote that he would choose to lead a scientific life, maintaining that "above all it is my individual disposition for abstract and mathematical thought, my lack of imagination and practical talent" that would determine his future. He wrote that, because he liked to do this kind of work, it was quite natural for him to choose a career in it, adding that he was attracted to the independence offered by the scientific profession. He was not, however, quite correct in this early self-appraisal. He would indeed prove to have some practical talent, and within the next 10 years he showed he had considerable imagination as well. In 1929, he was to write, "Imagination is more important than knowledge. For knowledge is limited, whereas imagination embraces the entire world, stimulating progress."

In the fall of 1896, Albert, after passing his final written and oral exams, was graduated from the Aarau school. On graduation, he had the highest overall marks in his class, receiving the best possible grades in mathematics, physics, and German in the compulsory essays. He impressed his examiners with the great talent he showed in these fields and with his simple yet clear and elegant writing. His worst marks were in French and geography.

At the end of October, Albert moved to Zurich to begin his higher education, sadly leaving the Winteler family behind, though he continued to think of Marie as his girlfriend for several months longer. But now his priority was immediately to enroll at the Federal Institute of Technology (later called the Federal Polytechnical Institute). Beginning in 1911, it became the Eidgenössische Technische Hochschule (ETH). This was the start of his lifelong love affair with academia and the time when he seriously began engaging in philosophical and other intellectually and emotionally satisfying discussions of all kinds.

Albert had gained entrance into one of the best teaching and research institutes in the sciences and engineering in Europe, one with a distinguished worldwide reputation and a faculty that attracted students from many nations. Here he was able to "seek out the paths that lead to the depths." Because his department did not formally prescribe a specific curriculum, his section head worked out a course of study for him. This program showed that Albert followed the plan for physics students very closely, and he abandoned his family's earlier recommendation to study engineering. All students at the ETH were required to take at least one class outside their main field of study. Albert, with his broad interests, enrolled in more classes than required, notably courses in philosophy, politics, and economics. Ready to start his life as an independent young man in pursuit of his professional goals, young Einstein rented a room on Unionstrasse in Zurich and settled in.

The budding young physicist was not an ideal college student, however. He was easily bored and somewhat arrogant, sharply criticizing those he considered to be of inferior intellect. He often skipped classes, preferring to study on his own in the library or to putter around in the physics labs. In physics, he felt he could intuitively sense the most important problems. Though he registered for at least nine mathematics courses, he did not attend all of them. Without shame, he borrowed mathematics lecture notes from those who had faithfully attended classes and were generous in sharing them, such as his classmate Marcel Grossmann. Einstein later wrote, "I would rather not speculate on what would have become of me without these notes." Grossmann, a mathematics student, would become a good friend and, later, a collaborator in Einstein's work on general relativity, for which he provided the complicated mathematical formulation.

Among the new friends Einstein made during his first semester at the ETH was fellow physics student Mileva Marić, the daughter of a well-to-do Serbian landowner and judicial clerk in Novi Sad, a town in what was at that time the Serbian part of Hungary. Albert admired this "older woman," four years his senior, for her intelligence and maturity. Slight and dark-haired, Mileva had come from Novi Sad to pursue her doctoral degree in physics in Switzerland, one of the few European countries that admitted women into its universities at that time. As a friend of hers wrote in 1898, Mileva was "a very good girl, clever and serious; she is small, frail, dark, ugly, talks like a real Novi Sad girl, limps a little bit, but has very nice manners." She shared Albert's passion for physics and his love of music, and it wasn't long before the teenager became completely enamored with her, writing her gushing love letters when they were apart. They often got together with friends to play music in small groups, with Mileva playing the piano or tamburitza, a mandolin-like instrument, and Albert the violin.

By the fall of 1899, Albert had already become eligible to apply for Swiss citizenship, and he now began this complicated process. He had much to do: fill out a long questionnaire; provide a statement of approval from his father, a birth certificate, a police report of good conduct, and a financial report; make a personal appearance before the naturalization commission; and, finally, pay fees to the municipality and canton of Zurich. Though he may have wanted Swiss nationality because he admired the nation's political system and institutions, his new status would also serve a more practical purpose. In a year or so, he might be looking for employment in the civil service sector, which included teaching positions, and these jobs required citizenship.

In the intervening year, Einstein continued his studies at the ETH. He chose his readings intelligently so he could remain on the cutting edge of physics, and he discussed his new insights with Mileva and Marcel Grossmann. He also spent much time in the physics lab, considering it the most time-consuming part of his studies. In his fourth year, he crammed for the final exams. He found exams distasteful and tedious throughout his earlier school years. Much later, he reflected on the topic of final exams in a short article he called "The Nightmare," which showed that exams must have been traumatic for him as well as for other students. In the essay, he proposed that the final exam students needed to pass before they could graduate from high school be abolished, contending that it didn't succeed in testing a student's knowledge and encouraged short-term rote learning. He felt the exam was useless as well as harmful—useless in that teachers can judge students better over the course of the many years they've taught them and harmful because students, concerned that their entire future rests on passing it, have nightmares preceding the day of the exam.

Nevertheless, Albert finished his exams at the ETH that final year without incident. He spent the summer vacationing with members of his family near Lake Lucerne. His mother did not lose any time questioning him about Mileva, whom she knew he had been dating. After Albert told her that Mileva would one day become his wife, Pauline feigned a nervous breakdown as her son looked on. Albert described the scene to Mileva:

"Mama threw herself down on the bed, buried her head in the pillow, and cried like a baby." Scenes similar to this one would continue over several years, but Albert dismissed his mother's histrionics and remained loyal to Mileva, surprising even his friends by his choice of this undistinguished and unhealthy if intelligent woman. To him, though, she was not only lovable but also a prospective colleague with whom he could discuss his life's work.

In the fall, Albert formally graduated with no particular distinction. Out of the five candidates in his section, he finished fourth. Mileva was fifth and failed to get a diploma. Despite his native intelligence, Albert had not been a favorite of the professors because of his lack of participation in the classroom and his cockiness. Professor Heinrich Weber, the head of the department of mathematics and technical physics and his supervisor, had also directed Mileva's program. Traditionally, when students graduated, the professors would hire them as assistants for a while, but neither Albert nor Mileva was offered a job. Weber and Einstein had a mutual dislike for each other, Einstein because Weber was a stereotypical German and Weber because he considered Einstein something of a rebel who had not distinguished himself in his studies. Mileva in particular held a grudge against Weber, especially after she failed her final exams.

Mileva went home to Hungary that summer of 1900 and stayed there for several months. She returned to Zurich in late fall, planning to take her exams again, and stayed and studied with Albert. By December she was anticipating that her sweetheart would soon have a job offer and fretted that he would be leaving Zurich soon to work elsewhere, complaining, "He is taking with him half my life." She went into a deep depression at the thought, even though Albert had no job prospects in sight.

As the twentieth century was born, scientists began to make technological and scientific advances that foreshadowed innumerable changes in our daily lives. They began to unlock the mysteries of the atom when German physicist Max Planck proposed that atoms emit or absorb radiation in packets, or quanta, launching the field of quantum physics. Also in 1900, Ernest Rutherford of New Zealand identified what he called gamma radiation, which consisted of electromagnetic photons rather than particles as in alpha and beta radiation. Sigmund Freud published *The Interpretation of Dreams* and launched inquiries into the mysteries of the mind, and Guglielmo Marconi was getting ready to send transatlantic signals from Cornwall, England, to St. John's in Newfoundland, Canada, the next year. This was only the first year of a remarkable new century in which scientific advances would dominate and change our lives to a greater degree than ever before.

Chapter 3

TO LOVE AND TO WORK

> I have decided the following about our future: I will look for a position *immediately*, no matter how humble it is. My scientific goals and my personal vanity will not prevent me from accepting even the most subordinate position.
>
> —To future wife, Mileva, July 7, 1901, while having difficulty finding his first job

In February 1901, Einstein had fulfilled the five-year waiting requirement and became a full-fledged citizen of "the most beautiful corner on Earth I know." Soon after he had received his Swiss citizenship, he was required to register for the military, an obligation he had wanted to avoid in Germany. Fortunately for Einstein, he was declared unfit for military service because he had some problems with his legs: flat feet, excessive foot perspiration (which may have led to his dislike of socks), and varicose veins. Instead of serving in the active armed forces, he was classified for inactive duty in the auxiliary military service. He never served even in this capacity and in lieu of it made annual military tax payments until the age of 42, as required by Swiss law.

Einstein was by now deeply involved with Mileva. Neither her limp—caused by childhood tuberculosis of the hip—nor her orthopedic boot or moody disposition deterred his feelings for the intense young woman. In the bloom of young love at the age of 22, after several years of courting her, Albert again stated his intention to marry her. Pauline Einstein still had an intense dislike for Mileva, though it is not clear why. Was it because of Mileva's disability? Because she was not Jewish? Because she was East European? Because she was four years older than Albert? Because, in Pauline's eyes, she wasn't pretty? Whatever the reason, she felt that Mileva was not good enough for her Albert, and she intended to do everything she could to keep him from marrying her. Her greatest worry was that Mileva might become pregnant. "What a mess you would be in then," she wrote to Albert, no doubt thinking about his lack of a job. Mileva was at first optimistic about winning Pauline's affections. In the end, though, when Pauline wrote a spiteful letter to Mileva's parents attacking her character, she learned that Pauline was obsessed with getting her away from Albert.

Albert had been having a difficult time obtaining favorable letters of recommendation. He was rejected for a teaching assistantship in Zurich, and he applied unsuccessfully for secondary-school teaching positions in the towns of Burgdorf and Frauenfeld. Finally, he settled on temporary teaching and tutoring work in other villages, existing on bare subsistence pay, all the time remaining in touch with Mileva in Zurich. He also wrote affectionate letters to her whenever she left Zurich to spend time with her family in Hungary. At the same time, he continued his desperate search for full-time and permanent employment by writing to physics professors all over Europe. Not one of them showed any interest in him, for which Einstein blamed Professor Weber.

Pauline's greatest fear was realized, for in the spring of 1901, Mileva became pregnant. The conception probably occurred during a romantic, secret rendezvous at Lake Como in northern Italy. Albert had just visited his parents in Milan and was about to start a two-month teaching stint at a technical school in the town of Winterthur. At first he seemed unperturbed by the impending blessed event but soon realized he had no means of support for the baby. He hesitated to get married before he had a job. As the baby's due date was approaching, he was still waiting to hear about a civil service position at the Swiss Patent Office in Bern for which he had applied half a year ago. He finished his temporary teaching job that summer and took a short vacation with his mother and sister while Mileva crammed for her final exams again. She failed for a second time and dropped out of her doctoral program. Disillusioned, depressed, and no doubt frustrated that her life was taking a direction she had not counted on, she returned home to Hungary to await the birth of her child.

Mileva made one short trip back to Switzerland in October to visit Albert, but both wanted to keep her presence a secret from others, especially from Pauline. By this time, Mileva's pregnancy must have shown—the baby was due around the end of January—so she stayed at a hotel in a village several miles away. Pauline was still obsessing about Mileva, maybe because she had learned about the pregnancy, which the young couple was keeping secret from their friends. In late 1901, Mileva, back in Hungary, hurt by the vehemence of Pauline's hatred of her, wrote to a friend that Mrs. Einstein "seems to have set as her life's goal to embitter as much as possible not only my life but that of her son. … I could not have thought it possible that such heartless and outright wicked people can exist!" Albert did not understand his mother's behavior, either, and continued to write supportive and loving letters to his "Dolly," his nickname for Mileva. She, in reply, called him "Johnnie." Both German forms were common terms of endearment in southern Germany.

Even though Einstein was having difficulties obtaining regular employment, he had not shied away from writing up the ideas he was developing on his own. For the work that interested him the most, he did not need a regular place of employment and a boss to tell him what to do and how to do it. "As a somewhat precocious young man, I was struck by the futility of the hopes and the endeavors that most men chase restlessly throughout life," he wrote later. "And I soon realized the cruelty of that chase, which in those days was more carefully disguised with hypocrisy and glittering words than it is today." As an independent scholar, he now began his own chase for fulfillment and submitted his first scientific paper, on capillarity (capillary action), to a prestigious German physics journal. The editors were impressed and published the paper of the unknown young scientist in March 1901, when he was just turning 22. Buoyed by his easy success, a few months later, in September, the young prospective father began work on a thesis on the molecular forces in gases that he hoped would lead to the Ph.D. degree from the University of Zurich.

Sometime in late January 1902, Einstein's first child, their daughter Lieserl, was born to Mileva at her parents' home near Novi Sad. Albert was not present. Through a letter that Mileva's father wrote to the new father after the baby's birth, Einstein learned that there were serious complications at delivery. The details are not known, but whatever they were, Albert expressed horror and sorrow at Mileva's suffering in a letter to her. Still, he did not go to see the new mother and daughter, perhaps for financial reasons. He had no money to spare, and certainly, under the circumstances, he couldn't ask his parents for the train fare. He wrote to Mileva lovingly, asking questions about the baby: "Is she healthy and does she cry properly?" "What are her eyes like?" "Which one of us does she look like?" "I love her so much even though I don't know her yet," and "I long for you every day." But Mileva remained in Hungary with the baby and her family for the next half year while Albert continued to write similar letters to her, planning their future together, telling her that he missed her desperately and asking questions about the baby. He did not mention marriage, no doubt because he was living in extreme poverty but also still afraid of his parents' probable reaction. In February, Pauline was still writing to friends: "We are resolutely against Albert's relationship with Fräulein Marić, and we don't ever wish to have anything to do with her. ... She is causing me the bitterest hours of my life." Mileva's parents, though they may have been socially embarrassed about the situation, were in a better position to take care of their daughter and granddaughter at this time than Albert was.

In February, perhaps energized by the birth of his child, Einstein packed up his meager belongings and moved to Bern in optimistic anticipation of an offer of work from the Swiss Patent Office. His friend Marcel Grossmann's father had already given him a favorable recommendation, and he had also managed to get support from his doctoral supervisor, Professor Alfred Kleiner in Zurich, who Einstein decided "was not quite as stupid as I'd thought." In January, around the time of Lieserl's birth and before he had moved, Einstein had abruptly withdrawn his doctoral dissertation. Kleiner may have had some reservations about the paper because it was critical of the work of another physicist, Ludwig Boltzmann, one of the inventors of statistical mechanics, and Einstein may have wanted to revise the thesis. This decision also gave him the freedom to move to another town.

Intent on making a living in Bern, the young father posted advertisements in the hope of attracting students whom he could tutor privately in mathematics and physics. A few responded, but not enough to enable him to earn the money he needed to get him out of his desperate financial situation. In the meantime, he made some new friends and reacquainted himself with old ones who had come to Bern to study. Among them was Paul Winteler, one of the brothers of his old girlfriend Marie, who had just begun to study law at the university. Another was Conrad Habicht, who was working on his doctorate in mathematics. And he made a close new friend, Maurice Solovine, who had answered his ad, wishing to be tutored in physics. The two young men often ended up discussing the philosophical foundations of physics as well.

Finally, in June, Albert got the break he was hoping for: the now legendary job for which he had applied at the federal Patent Office was his if he wanted it. The salary was still not high enough to support a family in an expensive city like Bern, but Albert took the job anyway. He became a low-level patent clerk examining mostly electrical gadgets and giving opinions on their usefulness. The probation period for the job would be about two years, after which the government office could either hire him permanently or, if he proved incompetent, fire him.

Mileva joined Albert in Switzerland sometime that summer or fall, inexplicably leaving baby Lieserl behind in Hungary. Albert had apparently still not seen his daughter. In the meantime, Albert's father became ill with heart disease from the stress of several business setbacks, and the situation soon became serious. In October 1902, the worried son rushed to Hermann's side in Milan. As he lay on his deathbed, Hermann told Albert that he no longer objected to his marriage to Mileva, then asked him to leave the room before he died. He was only 55, leaving Pauline a widow at the age of 44.

Less than three months later, on January 6, 1903, six and a half years after they met, a year after Lieserl's birth, and six months after he began his new job, Albert and Mileva were finally married in Bern. They had a simple civil ceremony with only two wedding guests—Einstein's friends Maurice Solovine and Conrad Habicht. The distraught Pauline, now mourning both the death of her husband and the marriage of her son to someone she detested, was not present. Young Einstein was the first in his family to have a non-Jewish spouse. The couple started their married life in a new apartment on Tillierstrasse in Bern, one of several they would inhabit during their seven years in the Swiss capital. Arriving home late after the wedding party, Einstein realized he had forgotten the key to their new home and had to wake up the landlord to let them in. He soon was able to report to his friends that he was leading a comfortable life with his new wife, that she was a good cook and housekeeper and always cheerful.

Bern was a lively and lovely town, with clusters of intellectuals and students seeking out one another in various cafés and homes. The newlyweds fitted right in. Einstein's need to express himself freely and mingle socially motivated him, Solovine, and Habicht to form, in the spring of 1903, what they playfully called their "Olympia Academy," after the mythical Greek gods who dwelled on Mount Olympus and had overthrown the Titans, the powerful family of giants who had ruled Earth. Their "academy" was a discussion group in which they informally talked about the important scientific and intellectual topics of the day. There they discussed the philosophical works of Karl Pearson, David Hume, Ernst Mach, Georg Friedrich Riemann, Baruch Spinoza, and Henri Poincaré, often well into the night. Sometimes Einstein would give a short violin recital or someone would read aloud from classical literature. These meetings also served as social occasions, at times boisterous ones. Mileva became part of the small group but remained more the quiet listener than an active participant, perhaps thinking more about baby Lieserl than physics and philosophy. Einstein later remembered these gatherings as filled with a "childlike joy" arising from "everything that was clear and intelligent."

Habicht, a fellow violinist whom Einstein had met the year before and who was now studying mathematics at the University of Bern, became a close friend and lifelong correspondent. Later in the decade, between 1907 and 1910, he, his younger brother Paul, and Einstein collaborated on what they called their "little machine," designed to detect and measure very small quantities of electrical energy. Solovine, whom Einstein affectionately called "Solo," was interested in the humanities and science.

In September 1903, while Mileva was visiting her family and Lieserl in Novi Sad, she sent Albert some bad news: Lieserl had come down with scarlet fever. Einstein wrote to Mileva expressing his concern for the child. He also asked how Lieserl had been registered, but Mileva's letter answering him has not been found and may have been destroyed. Even though Lieserl's birth or baptism must have been recorded somewhere, no evidence of either has ever been found. "Registering" may also have referred to adoption papers.

Einstein scholars have speculated for two decades about the reasons why Lieserl was not brought to Switzerland after her birth. It is possible that knowledge of the illegitimacy could have cost Einstein the civil service job he had been seeking at the Patent Office at the time of her birth. The parents may have decided to keep the birth a secret in Switzerland, which was socially conservative at the time, perhaps intending to bring her home after they were economically secure and married.

One thing is certain: Einstein, at least at first, had wanted and intended to keep Lieserl with them. He had written to Mileva not long after the birth that they still needed to decide how they could keep her. "I wouldn't want to have to give her up," he wrote to her in Hungary. Later, however, it seems that he made no effort to convince Mileva to bring the child home to them. Some have suggested that Lieserl may have had a medical condition or physical deformity, perhaps caused by the difficult birth, that prevented the parents from being able to take care of her. Most likely, however, she died of scarlet fever that fall—along with many others who had become ill with the disease—before her parents could make any permanent decisions about her care or adoption. In any case, around this time, according to her friends, Mileva became visibly distraught and moody, and they sensed that something of major consequence must have taken place between her and Albert. No doubt that, like any mother, she had wanted the baby to be with her, especially after the marriage. Mileva's family would probably have been in a position to offer temporary financial help. But, for whatever reasons, Albert resisted.

All that is known for sure about Lieserl is that after she contracted scarlet fever, she was never mentioned again in any of the surviving correspondence between either Albert and Mileva or anyone else. This does not mean that they and their relatives did not discuss or mourn her in their conversations, but no one in Bern seems to have known about her. Her existence remained a well-kept secret for decades. News of her birth did not surface until some 75 years later, after Albert and Mileva's letters dating from this period were discovered among the Einstein family correspondence. Lieserl's fate remains shrouded in mystery and continues to stump investigators.

Now, in the fall of 1903, a year and a half after Lieserl's birth and her possible death, Mileva wrote to Albert from Hungary that she was pregnant again. He wrote back saying he was happy to hear the news and eagerly awaited her return home. In late October, in anticipation of the baby's birth the following year, the couple moved into a modest two-room apartment on one of Bern's stylish streets. By this time, Mileva, after two failed final exams in physics and two pregnancies, seemed to have lost all interest in her own professional career. She embarked on her domestic role in life as a housewife, leaving the world of physics to her husband.

In December, Einstein gave his first lecture, "Theory of Electromagnetic Waves," to a broad-based scientific association of physicists, doctors, and others interested in the sciences he had joined in Bern. Although he attended meetings regularly, he did not often speak formally, preferring to give informal talks at the homes of other scientists. It was probably around this time that he also began to think about a new topic for the doctoral dissertation that he still wanted to submit to the University of Zurich.

On May 14, 1904, after another difficult pregnancy, Mileva gave birth to their first son, Hans Albert. This time the new baby stayed with the parents, and Einstein, at the age of 25, accepted the responsibilities of fatherhood and family. Until now, Mileva's parents, like Pauline, had not been happy about her marriage to Albert, but after the little boy's birth, they became more accepting. Pauline, too, resigned herself to the situation, and Albert and Mileva became more relaxed. Their marriage appeared to be happier, and the young couple settled into a family life that became emotionally and intellectually satisfying for both of them.

Einstein's professional life took a positive turn as well when, in September, the temporary appointment at the Patent Office in Bern became permanent. The patent clerk had passed his probationary period and received a small raise. The previous summer, he had encouraged his close friend Michele Besso, a young mechanical engineer, to apply also for a position at the Patent Office but at a higher level. Besso was thrilled that he was selected for the job from a field of 13 candidates, so now the two friends were firmly embedded with secure professions in the same city. Einstein had met Besso while both lived in Zurich and immediately liked him and respected his breadth of knowledge, eventually considering him "the best sounding board in Europe" for Einstein's own ideas. Just as important, he admired his friend for his fine personal qualities, showing great affection for him in their later correspondence. Besso had married Anna Winteler, the sister of Einstein's first girlfriend, Marie, in 1898. Anna and Mileva became good friends. Besso thanked Einstein later in life for being responsible for two important parts of his life that had brought him so much happiness: his wife and his job.

It was in Bern that Einstein, finally freed from economic worries, would spend the most creative years of his life. He conceived and developed his most revolutionary ideas during this period, mostly without benefit of close contact with the scientific masterminds of the time. In the beginning, however, according to Mileva's letter to a friend in March 1903, he had found that examining patents bored him to death, and he continued to look for a better job. But with Albert, seeking a job was always complicated, according to Mileva, because of his sharp tongue, and in anti-Semitic Europe, being a Jew didn't help. She wrote that both of them would even consider teaching German in Budapest, Hungary, if such positions became available—at this point she must still have had Lieserl on her mind and would have liked being closer to her or, best of all, having her live with them.

Even though his work in the office may have been disagreeable drudgery to him at the beginning, he was very much suited for it. The job required someone with critical judgment, able to ascertain the value or practicality of the many gadgets that passed through his office, making it important to think "out of the box." He was able to complete a whole day's work in just a few hours, giving him ample time to do some independent thinking. He was the only physicist in the group of mostly mechanical engineers, but he was there for good reason. At that time, many of the inventions in the rapidly developing electrical industry were based on a theoretical understanding of electromagnetic phenomena, with which Einstein was familiar. Even later, after he left the Patent Office and became famous, he was often asked for and gave his opinions on a number of patent applications. His willingness to do so was in line with his interest in technology and, according to him, also benefited his own research. During his tenure there, in 1906, a world-famous product received a patent in Bern, one that many of the clerks would no doubt have liked to have tested: the uniquely shaped Toblerone chocolate bar, the first patented milk chocolate with almonds and honey.

That Einstein was so unencumbered by the trappings and long hours of academic life and had the time, both in and out of the office, to think creatively paved the way for the brilliant work he was able to accomplish during his "year of miracles."

Chapter 4

THE PHYSICS OF THE "YEAR OF MIRACLES," 1905

> I've completely solved the problem. My solution was to analyze the concept of time. Time cannot be absolutely defined, and there is an inseparable relation between time and signal velocity.
>
> —In a letter to Einstein's friend Michele Besso, May 1905, telling him about his new theory of relativity

The year 1905, Einstein's *annus mirabilis*, or miracle year in terms of his contributions to physics, established Einstein, at the age of 26, as the world's leading physicist. Not only did he publish five important papers that year, but he also found time to write 23 review articles for a number of journals. He accomplished all of this on his own time after he came home from the Patent Office, a remarkable achievement for someone not yet stationed in the halls of academia. Despite his disconnection from academics at this time, he had no trouble making his way up the ladder in the physics world with his innovative if sometimes controversial contributions to the field. Not since Isaac Newton had his own *annus mirabilis* in 1666, during which he laid the foundation for his future work on mathematics, gravitation, and light, had any scientist made such significant contributions in one year.

As the year turned to 1905, Albert Einstein was 25 years old. He was a family man, with two years of married life behind him. His son, Hans Albert, was a toddler, and Einstein had a respectable though unexciting job as a patent clerk, third class, at the Swiss Patent Office. In the middle of this seemingly normal family life, Albert found time to think about physics. Perhaps he did so during the 20-minute walk between his apartment in the Gerichtesgarten and the Patent Office. He might also have worked on physics in the evening, after one-year-old Hans Albert had gone to bed, after they finished playing. In letters to friends, Albert mentioned that Mileva was helping him in his work, but we don't know how much influence she had on it. We know for sure, however, that they discussed his ideas, and no doubt she gave him feedback and helped him put his thoughts on paper.

No matter how he managed to concentrate on physics in his busy life, the results achieved in 1905 were spectacular. The five papers Einstein crafted that year rank among the most profound ever published in physics. One of them would finally earn him his Ph.D. degree and help establish that atoms truly existed. Two others launched a new area of physics—special relativity—for which he would become world famous. A fourth paper linked a curious observation about the erratic movement of pollen—Brownian motion—to the size of atoms. The last would earn Einstein the Nobel Prize for physics in 1921 and lay a foundation stone of modern physics—this was quantum theory.

Einstein was not a rookie at publishing. Five of his scientific papers had already appeared in print, the first one in 1901. All of them were published in the prestigious German journal *Annalen der Physik*. It is not clear that Einstein realized how provocative, if not miraculous, his work of 1905 would be. In May of that year, he wrote to his friend Conrad Habicht, describing the eventual Nobel Prize–winning work as a "paper that deals with radiation and the energy properties of light and is very revolutionary." Of his work on special relativity, Einstein said only that "the argument is amusing but for all I know the Lord might be laughing about it and leading me around by the nose."

* * *

Einstein completed his "very revolutionary" paper, "On a Heuristic Point of View concerning the Production and Transformation of Light," first. He submitted it to the *Annalen der Physik* on March 17, and it was published within three months. The paper explained the photoelectric effect, something that had puzzled scientists for almost 20 years, but accomplished far more: it laid a foundation for the new physics of quantum theory and challenged scientists to think of light as a collection of particles, or *light quanta*, rather than waves.

Ordinary white light, such as sunlight, splits up into its component colors under certain conditions. A droplet of water can do the trick, which is why—on a rainy day when the sun is still shining—rainbows sometimes appear. There's more to light, though, than the colors we can see. Visible light is only one example of what's called *electromagnetic radiation*. This phenomenon, it was always thought, traveled as a wave. Red light has the lowest energy and the longest wavelength of all visible light. *Infrared waves*, which we can't see, have slightly lower energy and longer wavelengths. Your TV's remote control, for example, uses infrared light to change the channels on your TV. If you want to detect infrared radiation, switch on an electric stove and place your hand above the hotplate. Even before the plate starts to glow red, your hand can feel the heat given off by the hotplate—this is infrared radiation. The shorter wavelengths below the infrared waves are *microwaves*, which can heat and cook food by making the water molecules in it vibrate rapidly. At longer wavelengths there are *radio waves*, which transmit radio signals.

It was the higher end of the spectrum, though, that caused problems for physicists in the late 1800s. At energies higher than violet light there is *ultraviolet light*, or UV radiation. Extended exposure to UV, as we now know, can cause sunburn and skin cancer. UV can also make something white really glow, so its use is popular in the advertising industry. An ultraviolet light shining on a white shirt or a pair of teeth in a TV ad will make them appear brilliantly white, as washing powder and toothpaste manufacturers know very well. Even higher-energy waves are the *X rays*. Not only do doctors and dentists use X rays, but out in space a black hole will generate X rays when it sucks gas in from a nearby star. Astronomers have used several satellites, such as the *Roentgen* and *Chandra*, to produce maps of the X-ray universe.

It was UV light that perplexed physicists. They knew that if it shone on the surface of some metals, such as tungsten, an electric current flowed on the surface of the metal. This is the *ultraviolet photoelectric effect*, and in 1905 no one could explain it, even though Wilhelm Wallachs had discovered it several years earlier, in 1888. A long article summarizing the photoelectric effect was published in 1904; perhaps Einstein read it and was intrigued by it.

In addition to photoelectricity, physicists had been unable to crack another problem regarding radiation. A hotplate, when switched on, is black. As it heats up, it glows a dull red, then becomes bright red. In days gone by, in a forge, a blacksmith could have heated up an iron rod even further, making it glow orange, then yellow, then a blue-white color. There was a connection, clearly, between how hot something is and what color light it emits, but what was it? At the dawn of the twentieth century, two rules were in place. Between 1900 and 1905, two English physicists, Lord Rayleigh and James Jeans, independently concocted a formula, the Rayleigh-Jeans law, to predict what intensity radiation would have, at a certain temperature, for the light of *long* wavelengths—the infrared, for example. Wilhelm Wien, who would win the Nobel Prize in physics in 1911, had a different formula that he had devised in 1896. He predicted the intensity of light at a particular temperature for light of *short* wavelengths, such as violet or ultraviolet. Heinrich Weber, who had been Einstein's teacher at the ETH and supervised his early Ph.D. dissertation work at the University of Zurich, had helped establish *Wien's law* experimentally. He obtained an experimental curve that predicted the wavelength of the most intense light at a given temperature, and in the winter of 1898–1899, he gave a series of lectures on the properties of radiation. Einstein was an undergraduate at that time and was enrolled in Weber's classes.

Physicists wanted a simple, single formula to explain the entire wavelength spectrum. German physicist Max Planck set to work on this project. Planck became one of Einstein's scientific heroes, of whom he would say, "How different, and how much better it would be for mankind, if there were more like him." In 1900, on or about October 7, Planck had an inspired idea. He wrote down an expression for how much energy a box of a certain size, full of radiation of a particular wavelength and temperature, would contain. He then did something completely different. He thought about an oscillator—a charged particle moving back and forth in a bath of electrical energy. How much energy does a collection of such oscillators possess? Planck answered that the energy of the oscillators and the energy of the radiation were the same. It was the key that unlocked the answer, for the oscillator problem was easier to solve than the radiation problem. He soon wrote down an expression that came to be called *Planck's law*, which did precisely what physicists wanted and needed. If you know the temperature and the wavelength of the radiation, Planck's law predicts what the intensity of the radiation will be. Astronomers put this formula to good use. Some stars, such as Vega in the constellation Lyra, are noticeably blue, whereas others, such as Betelgeuse in Orion, are obviously red. Now astronomers could measure the intensity of radiation emitted by a star or planet at certain wavelengths and thereby infer its temperature.

There was one unexpected problem, though. Planck could get his formula to agree excellently with experiment, but only if his oscillators at frequency f had a certain energy, E, for which $E = hf$. With this simple observation, quantum theory was born. From then on, photons—particles of light—could have an energy of only a certain specified amount, such as a multiple of hf. Light itself behaves like packets, or quanta, of energy.

Planck's formula appeared unquestionably correct. At short wavelengths, some terms in the equation became very small, and so Planck's law became Wien's law. At higher energies, other terms became tiny, and

Planck's law reduced to the Rayleigh-Jeans law. In addition, Planck's law agreed superbly with experiments if h, which eventually became known as *Planck's constant*, was roughly equal to 6.634×10^{-34} Js. Based on his value for h, Planck worked out the charge of an electron. He also calculated *Avogadro's number*, the number of molecules in "one gram mole of substance." Both values were important but poorly known in those early days of atomic theory. Planck's answer for Avogadro's constant, named after Italian scientist Lorenzo Romano Amedeo Carlo Avogadro, or Amedeo Avogadro for short, agreed with those of other experimenters, and Planck's electric-charge value remained the best for the next 10 years.

Einstein's paper on the photoelectric effect begins by showing that Planck was actually wrong. Einstein conceded that Planck's formula "agrees with all experiments to date," but he then had the temerity to take on one of the world's leading physicists and show that his reasoning was in error. A logical consequence of Planck's ideas, Einstein showed, was that energy would become infinite. In a brilliant move, Einstein showed that Planck's formula can still hold good. He does so by using a set of building blocks different from those used by Planck.

Einstein noticed a similarity between Wien's law and the well-known, widely accepted behavior of a perfect gas. By comparing radiation to a gas, Einstein came up with the *light-quantum hypothesis*: radiation behaves like energy quanta of energy $E = hf$. Einstein's reasoning was a triumphant blend of standard classical physics, cutting-edge experimental data, and a flash of genius. He had ended up with the same equation as Planck, but his logic was far more secure. But he did not stop there. He wanted to know: What are the consequences when light behaves like quanta?

For almost two centuries, scientists had regarded light as waves. An English physicist, Thomas Young, a man also proficient in several dead languages and a translator of Egyptian hieroglyphics, had devised a clever experiment. He passed light through a screen that had two slits in it. If light were made up of particles, Young reasoned, he would detect two streams of particles coming out of the slits. If light were made of waves, he would see patterns of light and dark, where the two waves emerging from the slit would reinforce each other or cancel out. This is the famous *Young's slit experiment*, and it proved that light has the form of waves. Nineteenth-century physicists were accustomed to thinking of light as waves, so they couldn't explain the photoelectric effect.

But what if light wasn't a wave? What if Isaac Newton, Max Planck, and Einstein were right, and light behaves like particles? In the ultraviolet photoelectric effect, particles of ultraviolet light hit the surface of a metal, causing an electric current to flow. What if the metal is made up of atoms? A brilliant idea came to Einstein's mind. A particle, a bullet of light called a *photon*, comes in with energy $E = hf$. The photon may hit an atom below the surface of the metal and knock loose an electron. Some of the photon's energy, called the *work function* P of the metal, is spent in getting the electron to the surface. The rest of the photon's energy sets the electron into motion; this moving electron is an *electric current*. This modern explanation lies at the heart of Einstein's explanation of the photoelectric effect. By pursuing the idea that light could be particles rather than waves, Einstein had solved the problem. He also showed that the size of the current does not depend on the intensity of the light shining onto the metal, confirming Philipp Lenard's experiments of 1902. Lenard, who won the Nobel Prize in physics in 1905 for his work on cathode rays, would later become a staunch Nazi and denounce relativity as "Jewish physics." American experimental physicist Robert Millikan confirmed Einstein's explanation of the photoelectric effect in 1916. Einstein's expression $E = hf - P$ for the energy of electrons emitted in the photoelectric effect allowed Millikan to carry out some high-precision experiments. Take a metal, shine light on it of frequency f, and detect the energy of the electrons emitted. A graph of E against f should be a straight line whose slope is h. Millikan reported that $h = 6.57 \times 10^{-34}$ Js to within an accuracy of 0.5 percent.

* * *

To obtain a Ph.D. degree, a student works closely with an adviser to find an important unsolved problem to study. The student solves the problem, with little or no help from the adviser, and writes an extended report, the Ph.D. thesis or dissertation. The student presents the thesis to examiners who judge whether the work is original, important, and correct and merits publication in a good physics journal. Though Einstein had originally submitted a thesis in November 1901, it was not accepted. Somewhat despondent, he wrote to his friend Michele Besso in January 1903 that he had given up on getting a doctorate, as "it doesn't help much and the whole comedy has become tiresome." Einstein would change his mind. Part of the problem might have been due to his thesis adviser, Heinrich Weber, with whom he did not get along well. Einstein subsequently relied more heavily for advice on another professor, Alfred Kleiner, under whose direction he completed his new thesis, "A New Determination of Molecular Dimensions," on April 30, 1905. According to legend, Kleiner told Einstein that the thesis was too short. Einstein added a single sentence and resubmitted it. It was sent to the university authorities on July 20, and the Ph.D. degree was awarded soon after. Einstein dedicated the thesis to his good friend Marcel Grossman, the mathematician who would play a key role in the development of the general theory of relativity.

Physicists had only recently become used to the idea of atoms. In the early 1900s, there was still a dearth of data on the size of atoms and on how many atoms were contained in a given amount of substance, Avogadro's number. Einstein, in his Ph.D. thesis, came up with a new way to determine both. To do so, he looked at the flow of a liquid. Think of a river flowing downhill. If you place your hand in the river, the flow of the water will change. Einstein looked at how the flow of a liquid changes if you place a small sphere into it. He then generalized upward to find out what happens when there are N spheres in the liquid rather than just one. At least two things occur. First, the viscosity of the liquid changes. The presence of the spheres creates a new "effective" viscosity that is higher than the natural viscosity of the liquid. Second, the spheres diffuse slowly through the liquid; that is, the spheres move slowly through it, spreading out as they go. Physicists use a single number, D, the diffusivity, to account for how rapidly the spheres spread. Einstein calculated what the new viscosity would be in terms of the radius P and the number of the spheres; he then calculated the diffusivity in terms of P and N. Cleverly, Einstein changed these equations around to find out what P and N are in terms of D and the viscosity. This was crucial because viscosity and diffusivity can easily be measured in simple "tabletop" experiments. Einstein then plugged in values for the viscosity and diffusivity of a solution of sugar dissolved in water. With these data, Einstein estimated that an atom has a radius of about 10^{-9} meters. He also estimated that there are about 2×10^{23} atoms in one gram-molecule of a substance, the quantity now known as Avogadro's number. These were in excellent agreement with other methods used in his day.

With his thesis submitted, Albert and Mileva took young Hans Albert to Belgrade and Novi Sad to meet his maternal grandparents, Milos Marić and Marija Ruzi Marić.

* * *

It would not be long, only 11 days, before Einstein sent off another important paper, "On the Motion of Small Particles Suspended in Liquids at Rest Required by the Molecular-Kinetic Theory of Heat." It, too, appeared in the *Annalen der Physik*, and it would become the Einstein paper most often quoted by other scientists. In it, Einstein began by using a slightly different method to calculate the diffusion coefficient from the one he employed in his Ph.D. thesis. He then looked at a cloud of molecules to determine how they spread out in a liquid. If a molecule is at a point X

at one instant, where will it be at the next instant? Einstein's answer involves some complicated mathematics, but the basic idea is simple.

Suppose you toss a coin. If it comes down heads, take a step to the left; if it comes down tails, take a step to the right. If it's a fair coin, then on average the number of steps to the left and right will even out, and on average you will be at the place you started. Suppose you toss the coin four times, and it comes down heads, tails, tails, heads. After one throw, you're one step to the left; after two, you're at zero; after three, you're one step to the right of where you started; and after four, you're back home. On average, you remain at the position where you started. What if we don't care about whether we are left or right of the starting point? That is, what if we care only about how many steps we are from our starting point? In the example, after one toss, we're one step away; after two steps, we're zero steps away; after three steps, we're one step away; and after four steps, we're back home. The average of these is one-half step. Physicists usually find it easier to square the distances and therefore report the so-called *mean-square distance*. After tossing the coin N times, the mean square distance is proportional to N. It turns out that atoms behave rather like the person hopping to the toss of a coin. In each small period of time, an atom moves a certain distance, and on average it gets nowhere. The mean-square distance depends on the number of steps taken, which in turn depends on the time that has elapsed. Einstein realized a couple of things. First, the atoms would spread out in this strange "random walk." Second, the spread of particles that have this strange, random motion was the same as the spread caused by simple diffusion. Most important, he could find a formula to express the average of the squares of the distance traveled by a particle to the time t. The slope he got would allow physicists another method by which to determine Avogadro's constant and the size of molecules.

His prediction helped explain the observations of the Scottish biologist Robert Brown in 1828. Brown had noticed that pollen moved in zigzag, jagged paths. Some biologists in the nineteenth century even believed that the particles were alive. When that idea was ruled out, some thought electricity might be responsible. Einstein, though, had presented a clear picture: air molecules, in their random motion, buffet the pollen grains. A Polish theoretician, Marian Ritter von Smolan Smoluchowski, confirmed Einstein's theory in 1906, using perhaps a more elegant approach than Einstein. Their joint theory cried out for an experiment, but these were extraordinarily difficult to perform. Within a couple of years, Einstein would write articles to explain this theory of Brownian motion to chemists who were extremely interested in the phenomenon but who lacked the math background that physicists have.

In 1908, a French experimenter named Jean Perrin confirmed the predictions. For this work, Perrin would win the Nobel Prize for 1926, five years after Einstein. It certainly was well deserved. Instead of using pollen, Perrin prepared thousands of tiny particles of gamboge (a Southeast Asian tree resin) and mastic (a resin from mastic trees). He spun these in a centrifuge, a rapidly rotating machine, and skimmed off small particles of roughly the same size. These he placed under a microscope and patiently watched and recorded how they moved—for hours. His results agreed beautifully with Einstein's theory. Perrin went even farther and managed to determine experimentally the rotational motion of the gamboge and mastic. Perrin's results surprised even Einstein, for although in 1906 he calculated the effect of rotational Brownian motion, he never thought of it as more than a "pretty little idea."

* * *

It was for the theory of relativity, however, that Einstein became famous. Here, too, he latched on to a controversial idea and worked out its consequences. In the middle of the 1600s, English scientist Sir Isaac Newton had come up with three equations of motion. These, combined with Newton's calculus, could predict the movement of objects acting under the influence of simple forces. His laws of motion were spectacularly successful and could be used to predict not only how a ball might move through the air but also how planets move around the sun. There was no question, in the middle of the nineteenth century, that Newton's laws were correct. By the end of the nineteenth century, however, some physicists had begun to have nagging doubts about the laws, even though most of them still believed that almost all the major laws of physics had been discovered—that the only task that lay ahead was to clean up the loose ends. They were wrong.

In 1864, Scottish scientist James Clerk Maxwell combined electricity and magnetism into one elegant unified force, *electromagnetism*. His four equations of electromagnetism made a clear prediction: light was an electromagnetic wave and would travel at a constant speed. This created a problem. Suppose you are in a car traveling at 30 miles per hour, and a car going at 55 miles per hour overtakes you. Sitting in your car, the other car seems to be going away from you at 25 miles per hour. That seems simple and straightforward. The relative speed is merely the difference in the two speeds, $55 - 30 = 25$. But what if the car has its headlights on? Suppose the speed of light, relative to the driver of the other car, is c. Then if you measure the speed of the light emitted by the headlights of the other car, you should get an answer of $55 + c - 30 = c + 25$. The only way for c to be the same as $c + 25$ is if the speed of light is infinite. That's not what Maxwell predicted, however. Physicists were in a quandary: Maxwell *or* Newton could be right, but not both of them. Given the 250-year track record of Newton's laws, most physicists thought Newton must be right. But Einstein backed Maxwell.

Attempts to measure the speed of light went back many centuries. If you're in a large, mostly empty room and speak, chances are you'll hear an echo. Your larynx, the voice box, has produced sound waves that travel to the other end of the room and then bounce back. You hear these reflected sound waves a fraction of a second after you have spoken. If you know how far away the wall is and measure the time delay between speaking and hearing the echo, you'll know how fast the speed of sound is. Galileo, in his book *Dialogue on the Two New Sciences*, published in 1638, suggested a similar experiment, but with light. Two people stand far away from each other. Each has a covered lantern. The first person uncovers his lantern, which sends a beam of light to the other person. When the second person sees the beam of light, she uncovers her lantern, sending a beam of light back to the first person. If the people know how far apart they are and measure the time taken for the first person to see the beam of light from the second, then they can work out the speed of light. It's not clear if Galileo ever did the experiment himself, but he certainly knew that the speed of light was very fast indeed, for Galileo had observed that lightning appears well before thunder and that the flash from a cannon arrives before the sound of the gun being fired. So he knew that the speed of light was far higher than that of sound.

Galileo, with his telescope, was the first person to observe closely the moons of Jupiter. The first major attempt to determine the speed of light was by the Danish scientist Ole Roemer in 1675. He looked closely at Jupiter's moons and noticed a tiny time delay between when the moons *should* go into eclipse and when they *actually* did so. This delay, he thought, must be due to the finite speed of light, the time taken for the light to get from Jupiter to us. Roemer worked out that light took about 11 minutes to travel from the Sun to Earth, but he didn't know the size of Earth's orbit. In 1669, French astronomer Jean Picard had found a value for the diameter of the Earth. A French-led expedition to Guyana enabled astronomers to work out the distance to the Sun in terms of the Earth's diameter. Combining all three observations allowed scientists to calculate the speed of light. The first to see his calculation in print was Dutchman Christian Huygens, who wrote in his 1690 book *Treatise on Light* that its speed was 130,000 miles per second.

James Bradley, in 1726, also took an astronomical approach. Stars appear to move across the sky, and the angle at which they appear depends on the speed of light and the speed of the Earth's orbit. Running in the rain is rather similar: the angle at which the rain hits you depends on your speed and on the speed of the droplet. Roemer and Bradley, who would become Astronomer Royal of Great Britain after Edmund Halley of Halley's Comet fame died, got good estimates for the speed of light.

In the 1850s, two French physicists, Armand Fizeau and Leon Foucault, devised ingenious laboratory-based experiments to find the speed of light. Foucault's method was arguably better, but he suffered from having a lab that was too small to conduct a good experiment. Not until 1924 did Polish-born American Albert Michelson use a 22-mile outdoor setup in California that would obtain experimental results showing that the speed of light is around 186,000 miles per second, or close to 300,000 kilometers per second.

If light is a wave, what does it travel through? Sound waves travel through air or water. Ripples on a pond are waves of water traveling through air. Light, surely, had to travel through something. This unknown "something" was what physicists back then called the *ether*. Many theories were proposed for the ether, but none of them stood up to experimental tests. Three great ether experts—Lord Kelvin, Lord Rayleigh, and Albert Michelson—got together at the Johns Hopkins University in Baltimore in 1884. Three years earlier, Michelson, then a master in the U.S. Navy serving at the Naval Academy in Annapolis, Maryland, had completed an experiment that seemed to rule out the ether. Rayleigh urged Michelson to perform the experiment again, and so he joined forces with an American chemist, Edward Morley, to refine the experiment that marked the end of the ether hypothesis. Einstein, in 1905, showed that the ether hypothesis simply wasn't needed. Michelson won the Nobel Prize in physics in 1907.

The *Annalen der Physik* received Einstein's first paper on special relativity, "On the Electrodynamics of Moving Bodies," on June 30. He had been working on what later came to be known as the special theory of relativity at least since 1899 and had finally completed formulating his ideas. In the paper, Einstein showed what happens if the speed of light is constant. Instead of using cars—which weren't common in 1905—or trains, he came up with the idea of a reference frame. If one reference frame (say, a train) travels at a speed V relative to another reference frame, what happens if the speed of light, c, is the same in both frames? Weird and bizarre things occur. Space and time are no longer separate but become welded together into a single item, *spacetime*. This may seem odd, but we do similar things every day. Ask someone how far it is from Baltimore to Washington, D.C., and the answer you get might be "45 minutes": your question asks for a distance, but the answer you get is a time. The two are linked, if you obey traffic laws, by the speed limit. In the universe, distance and time are connected by the ever-constant speed of light.

If c is indeed the same—186,000 miles per second—in every reference frame, some odd behavior occurs, according to Einstein. If one reference frame is at rest while the other frame moves at a constant speed V, then a rod in the moving frame is shorter (according to someone in the stationary frame) than the same rod measured in the frame at rest. This is *length contraction*. The theory had been worked out earlier by Dutch physicist Hendrik A. Lorentz and, in 1889, by an Irish physicist, George Fitzgerald, whose "brain was too fertile and inventive," according to fellow eccentric Oliver Heaviside. Lorentz and Fitzgerald had originally come up with the formula to help explain the unexpected result of the Michelson-Morley experiment in an effort to save the ether. While the ether theory has long since passed away, the Lorentz-Fitzgerald contraction lives on.

If you are in a reference frame "at rest," a measuring rod (a stick or a ruler, say) in a moving reference frame seems to be length contracted. You think its length is shorter than the length measured by your friend who sits in the "moving" frame. The curious thing is that your friend doesn't think she's in a moving frame at all—she thinks *she's* at rest and *you* are moving relative to her. In her opinion, the rod in your reference frame is shorter than hers. Both of you are correct. Neither you nor your friend has a special frame of reference, or a preferred view of the world: *everything is relative*.

In a similar way, a clock in the moving frame runs more slowly compared to the clock measured in the rest frame. This is *time dilation*. Again, a clock in your friend's "moving" frame runs slowly compared to the one in your reference frame; but your friend thinks your clock runs more slowly compared to hers. Time dilation leads to the famous *twin paradox*. Separate twin siblings at birth and whisk one away at speed V and then eventually reunite them. Is the traveling sibling younger than the one who stayed home? Scientists put this to the test using some very precise atomic clocks. One clock remained in the laboratory, and the other was transported by a high-speed plane that flew in a "race-course" path. The two clocks were compared, and, lo and behold, the clock that had been on the plane was slower than the one that remained in the lab. This might seem to be a paradox. In special relativity, there should be no difference in the twins' ages. One twin thinks that her brother has gone off at speed V and returned. The brother, though, thinks his sister went off and returned. In that case, both twins should have the same readings on their clocks. In the experiment, the laboratory clock and the airplane clock should show the same time, but they don't. There is no paradox, though. Special relativity deals only with constant velocities, and for one clock or one twin to go away and come back, there must have been some acceleration or deceleration of the clock or twin. So, the situation is not covered by special relativity at all, and all bets are off.

Both length contraction and time dilation are confirmed by highly accurate experiments. Radioactive particles in a laboratory decay away after a certain time, the *half-life*, and the amount of "parent" radioactive particles can be measured, as can the number of "daughter" particles, that is, the nuclei into which the original particles decay. If the particles travel in a beam relative to the laboratory, time is dilated for the particles in the beam. This means the particles can go farther and live longer than we, in our rest frame, think they should. That's precisely what the experiments repeatedly show, which is proof of the bizarre predictions of time dilation and length contraction.

Einstein also showed that if a light shines in a frame receding at speed V, it appears redder than a similar light source in the rest frame. This is the relativistic *Doppler shift*, named after the nineteenth-century Dutch physicist Christian Doppler. Doppler showed how frequencies change when sound comes out from a moving source. A good example is the drop in the pitch of a police-car siren when it overtakes you. Einstein predicted a similar shift in the frequency of light coming from a source that moves close to the speed of light.

Three months after the first relativity article, in September, Einstein sent off a follow-up paper. In "Does the Inertia of a Body Depend on Its Energy Content?" Einstein studied the kinetic energy of a moving body. When the speed V of the moving reference frame was low, Einstein's formula for the kinetic energy matched Isaac Newton's, but with one critical difference. When the speed V is zero, an object of mass m still has a nonzero energy, E. This is the *rest-mass energy* of the object. Although he did not write the equation down until 1907, Einstein's formula for the rest-mass energy would become famous the world over as $E = mc^2$. In 1905, he published papers that strongly suggested that light behaved like particles and that the light particles had restricted energies. In 1907, he came up with an equation that, in hindsight, showed that photons had a definite momentum. In his second 1905 paper on relativity, Einstein indicated that the rest-mass energy implies that energy and mass are equivalent, which might explain radioactivity. As he put it, you might "test this theory using bodies whose energy content is variable to a high degree (e.g., radium salts). If the theory agrees with the facts, then radiation carries inertia between emitting and absorbing bodies." Little did he realize that this observation and his simple equation would profoundly alter his life, for it inadvertently led to the construction of the atom bomb some 40 years later.

Chapter 5

ACADEMIC LIFE IN SWITZERLAND

> Freedom of teaching and of opinion in books or in the press is the foundation for the sound and natural development of any people.
>
> —From a talk delivered in 1936 to the organization Freedom of Opinion

While working at the Patent Office and still looking for better work, Einstein had decided to dispense with getting a doctoral degree, feeling it wouldn't help his career. But the lure, prestige, and benefits of an academic life were apparently still attractive to him, so, in the summer of 1905, he renewed his pursuit of the Ph.D. degree. It would be a simple matter—he already had some works in progress that he could finish and submit. At first he submitted his paper on relativity, "On the Electrodynamics of Moving Bodies." This paper, which would later make him famous, was rejected by the dissertation committee, so on July 20 he submitted another one, "A New Determination of Molecular Dimensions," which was more to the committee's liking. They read the 17 pages quickly, praising the work as being of high quality and showing great insight into physics. Their only criticism was that it lacked detail and had "stylistic infelicities and slips of the pen." The whole approval process was completed in August, and Einstein quickly became a "Herr Doktor," even though he would not receive his degree formally until the following January.

Two months later, in March 1906, with the coveted doctor of philosophy degree in hand and several important publications under his belt, Einstein was promoted to a higher-rank patent clerk in the Patent Office in recognition of his ability to analyze difficult patent applications. With the promotion came the higher salary Einstein had been hoping for, and the young family soon moved into a larger apartment. But the papers he had written the previous year had not yet made the rounds of the scientific community, and he had received little recognition for them right after publication. Certainly, the leading physicists must have been discussing his work, even though he may not have been aware of it, for it had been published in the prestigious German journal *Annalen der Physik*.

Now, after his year of miracles, Einstein eagerly waited for reactions to his publications. Until now, he had been a veritable outsider in the world of physics, unknown for his hard work except by those close to him. But things were about to change. Some of the most eminent physicists of the day, such as the Dutch Nobel Prize winner H. A. Lorentz and future Nobelist Max Planck of Germany, soon took notice, understanding and appreciating the revolutionary implications of Einstein's work. He began a voluminous scientific correspondence with many of the European physics luminaries of the day.

Einstein had warm regard for Lorentz, who soon became a father substitute for him. He sent long and venerating letters to the older man, a few years later writing to a friend, "Lorentz is a marvel of intelligence and exquisite tact. A living work of art!" To Lorentz himself he wrote, "My feeling of intellectual inferiority with regard to you cannot spoil the great delight of our conversations, especially because the fatherly kindness you show to all people allows no feelings of despondency to arise." Lorentz was famous in Europe, having already suggested earlier, before Einstein postulated his own relativity theory, that a moving object might appear to contract relative to a stationary observer, and Einstein provided a logical explanation of why this should happen.

Einstein was also a great admirer of Planck's, who was the co-editor of the *Annalen der Physik*. He later called him "one of the finest persons I have ever known," even after Einstein spoke out against other German scientists during the Hitler years. A few years earlier, Planck had completed work suggesting that radiation is emitted from objects in small bursts, called quanta, answering the old question of how a hot object radiates its heat at different temperatures, for which he would win the Nobel Prize in 1918. Einstein credited Planck's support of his relativity theory for attracting the notice of other colleagues in the field so quickly. He remained close to Planck in later years, consoling the German physicist as he endured a succession of dreadful losses and suffered the consequences of two world wars: his first wife died in 1909, his elder son was killed in action in 1916 during World War I, and his daughter died after childbirth the following year. As if this were not enough, Planck's house was burned down in an air raid in World War II, and his other son was executed in 1945 after he admitted to taking part in a plot to assassinate Hitler.

These prominent men and many others now sought out the young physicist for scientific discussions, and Einstein expanded his circle of colleagues throughout Europe. For the next few years, in order to convince his critics of the validity of his theory, he concentrated on publishing a number of papers to elaborate his ideas. But it was not until almost a quarter of a century later, during the early 1930s, that his special relativity theory could be confirmed experimentally, with the study of nuclear reactions, and time dilation was not directly proved until 1938. This was an unusually long time for the confirmation of a theory.

Mileva, meanwhile, felt she was being pushed more into the background as physics became increasingly important to her husband's career and happiness. Still, Einstein was an attentive family man at this time and especially doted on his little Hans Albert. To mitigate her loneliness, Mileva stayed in touch with her family and took the little boy to Hungary to spend some time with her parents. They showered each other with affectionate letters in which Einstein told Mileva about his latest scientific triumphs and Mileva told him about her time with her family. The couple's marital relationship appeared sound.

By this time, the small but lively Olympia Academy had disbanded because its members had moved to distant places. Einstein especially missed his good friend Solo, who had moved to France and became a translator. But he now had more time to spend with his little son, Hans Albert, whom he called Albert or "Adu." He proudly reported that Albert was already as impertinent as he and was giving him and Mileva great pleasure with his precociousness. With the demise of the Olympia group, Einstein now took the opportunity to deepen his friendships with other Bernese residents, such as Michele Besso and his sister Maja's fiancé, Paul Winteler. Maja had also moved to Bern that year to continue her studies for the Ph.D. degree in Romance languages. She accomplished this feat in 1909 after writing a dissertation on an Old French manuscript about the Chevalier au Cygne, a mysterious knight who arrives on a boat pulled by a swan *(cygne)* and rescues a damsel from an evil knight. This accomplishment was quite a feat for a woman in those days, and now Pauline was able to boast that both of her children held the esteemed "Doctor" title. In 1910, Maja and Paul, who had received a law degree in Bern, were married.

Partly based on a letter from Einstein to Mileva as early as March 1901, in which he referred to "our work," some historians have suggested that Mileva was Einstein's collaborator in his scientific research. Though this claim cannot be ruled out entirely, it cannot be substantiated, either. From their correspondence when they were apart, it is known that Albert would run his ideas by Mileva, and she would take interest and make suggestions. It is also likely that she proofread his papers and pointed out inconsistencies. But physicists who have evaluated Mileva's background in Einstein's field of concentration think it unlikely that she was a creative force behind his work. Furthermore, at the time Einstein was making his most important contributions, Mileva was a new mother and housewife, and even in her letters to friends she made no mention of current intellectual undertakings. Nevertheless, as a physicist herself, she surely would have had an interest in her husband's work, and all agree she should receive credit for any help, input, and discussions that may have helped her husband formulate his ideas.

In his papers, Einstein did not provide a clear history of the origin and precursors of relativity theory, so it is uncertain what ideas influenced him directly. It is likely that French mathematician Henri Poincaré, who had written on "The Relativity of Space" in 1897, and H.A. Lorentz, who in 1895 published "Inquiry into a Theory of Electrical and Optical Phenomena in Moving Bodies," helped set Einstein's ideas in motion. Many physicists believe these two men contributed greatly to relativity theory and to the path that led Einstein to his own revolutionary conclusions. In 1921, when Einstein visited America, he said, "The four men who laid the foundations of physics on which I have been able to construct my theory are Galileo, Newton, Maxwell, and Lorentz." James Clerk Maxwell had laid the groundwork in electrodynamic, electromagnetic, and radiation theory in the late nineteenth century. It is not clear why Einstein did not mention Henri Poincaré's early work, which also culminated in Poincaré publishing his own ideas on a relativity principle in 1905 and 1906. On the other hand, Poincaré also ignored Einstein's work and stuck to his own concept of relativity. In any case, the time was ripe in 1905 for the development of relativity theory, and Einstein later expressed his belief that if he had not made the discovery in 1905, someone else would soon have done so. He may have been fortunate in having more time to think creatively and make the necessary connections to formulate new ideas while spending his workday at the Patent Office: he was at the right place at the right time—but his creativity and intelligence played an equally large if not larger role.

Einstein's work, especially his studies on relativity, was now being discussed seriously by the most prominent physicists in Europe. These men were unaware the young genius was a clerk in the Patent Office in Bern rather than a professor at an acclaimed research institution. Despite his growing reputation and his having published 25 papers in three years, he was still having a hard time obtaining the academic position he longed for. Much of the problem was due to his obstinacy in not following established rules. For example, in June 1907, he applied for a postdoctoral position at the University of Bern, but the university's officials rejected his application because he had not fulfilled one of their requirements: a new, unpublished thesis that showed he was capable of undertaking original research. In its place, he had sent the search committee a bundle of his previously published papers, his personal opinion being that these should suffice as a precondition for the job. Also enclosed were his diploma, dissertation, and a résumé consisting of nine lines. After his offerings were rejected, he unsuccessfully tried to obtain a position at the Gymnasium in Zurich and then thought about applying for a position at the technical college in Winterthur, where he had worked temporarily in early 1901.

In the fall of 1907, after contemplating why relativity seemed to apply to almost every physical phenomenon except gravity, Einstein had an exceptional flash of genius he called "the happiest thought of my life." He formulated the "principle of equivalence for uniformly accelerated mechanical systems," which showed that gravity and acceleration are directly equivalent. The 1905 relativity paper had not shown how to deal with the question of acceleration. While sitting in the Patent Office, the idea suddenly occurred to him that if a person were in a free fall, he would not feel his own weight. Because everything falling with the person falls at the same rate, there is no way to tell he is in a gravitational field—there is no reference point. Einstein concluded the person could assume he and everything immediately around him were at rest, while all the objects farther away that he could see were being pulled upward. In other words, for Einstein, gravity seemed to be relative. This observation is not so different from Isaac Newton's thought that the same kind of force that pulled an apple to the ground must also hold the moon in orbit. In Newton's case, the development of his idea eventually led to his famous work *Principia Mathematica;* in Einstein's case, it would lead to general relativity theory several years later, which would enlarge the more limited special theory. During this time, Einstein also became interested in the unexplained motions of the planet Mercury.

In January 1908, Einstein faced the reality necessary to make one's way into established institutions: you need to follow the rules, at least at the start. He finally submitted the required thesis, titled "Consequences for the Constitution of Radiation of the Energy Distribution Law of Black Body Radiation" (which remains unpublished because he apparently threw it out at some point), that would qualify him for a part-time lecturer's position at the University of Bern. The faculty moved quickly and hired him the month after its submission, and he delivered the traditional inaugural lecture at the university at the end of February. His university teaching career began in late April with a course on the molecular theory of heat. Because the salary was so paltry, he continued to work in the Patent Office as well. Mileva wrote to a friend about the teaching position: "Unfortunately, this post is so miserably paid that we cannot even feel happy about the honor." After the course ended in late July, he headed to the Alps for a vacation with Mileva and Hans Albert. During the winter semester late that year, he began to teach a course on the theory of radiation.

Throughout 1908, Einstein, along with his friends the Habicht brothers Paul and Conrad, was intensely involved with the continuing experiments on their "little machine," designed to measure small amounts of energy. Paul Habicht had just started a small company that made scientific instruments, and he set about building the device. After its completion, they tried to get a patent for it but dropped the idea because manufacturers showed no interest in it at the time. Instead, Einstein published an article to describe the principles of the little machine. In time, the gadget proved useful because it could measure radioactivity and test relativity theory by measuring the equivalence of mass and energy. They eventually did receive a patent, but the instrument never became popular, and only a few were ever manufactured.

Most people think of Einstein as a theoretical physicist, but besides working on this little machine, for many years, he also enjoyed tinkering with gadgets, solving complicated puzzles, and experimenting in labs. Both his son Hans Albert and his granddaughter Evelyn recalled in later years that Einstein was an enthusiastic tinkerer. Hans Albert's favorite toy was a cable car his father had built from matchboxes, and Evelyn remembers that her grandfather played with complicated puzzles that could be taken apart and put together again, often given to him by friends. (See chapter 12 for information on Einstein's experiments and patents.)

Einstein's reputation as a young physicist in his prime, full of innovative ideas, continued to grow and spread in 1909. His desire for a more important position was soon to be fulfilled. His old mentor in Zurich, Alfred Kleiner, whom Einstein had earlier declared "not so stupid after all," was pushing his department at the University of Zurich to appoint a theoretical physicist and was considering offering the job to Einstein. He went to Bern to evaluate Einstein's teaching skills and was not impressed. Einstein agreed he had had a bad day because he was not prepared and admitted he had been nervous about having Kleiner in the audience. Kleiner decided not to make the offer but then reconsidered, giving Einstein another chance to prove himself. He invited him to lecture to the Physical Society of Zurich in February 1909, and all went well. Kleiner was able to convince his search committee that Einstein was one of the most important theoretical physicists of the day, one whose publications showed a profound understanding of physics. He wrote to his colleagues, "Today Einstein ranks among the most important theoretical physicists." The committee soon voted to appoint him to a position as associate professor of theoretical physics, to begin in October 1909. It was Einstein's first job as a full-time professor.

Mileva was proud of Albert and their new status but had mixed feelings, too, knowing that with this anchor in academia her husband would be free from the predictable eight-hour routine of his Patent Office job. "He will now be able to devote himself to his beloved science, and only his science," she prophetically proclaimed to a friend. Albert's starting salary at Zurich, however, would be the same as the salary he had been earning as a patent clerk, a situation he hoped would soon change.

During the summer of 1909, on hearing that his appointment came through, Einstein submitted his resignation from the Patent Office, his professional home for the past seven years. Two days later, the new professor traveled to Geneva to receive his first honorary doctorate from its university. (In his lifetime he would receive approximately 25 such honors.) In fact, he almost didn't make it to the celebrations. He had tossed out the letter informing him of the honor and inviting him to come, thinking it was a piece of junk mail. After he hadn't responded, the university's officials persuaded him to come to the city, though he still didn't understand why. When he was finally told, he couldn't take the occasion very seriously. He was the only man to arrive in a summer suit and straw hat while other dignitaries and recipients showed up in their finest clothes or academic robes. Another recipient was Marie Curie, and it must have been here where the two scientists first met, though most biographies claim they made their initial acquaintance at the first Solvay Congress in the fall of 1911. Another honoree on this day was Nobel Prize–winning chemist Friedrich Ostwald, who would later nominate Einstein several times for the Nobel Prize in physics. The times had changed, for only a few years earlier, in 1901, both Einstein and his father, each without the other's knowledge, had sent letters to Ostwald asking if he could help find work for the jobless young Einstein.

Mileva was hesitant to leave Bern that fall. She became increasingly insecure about Albert's fame and its effect on the family. He was often called away from her and Hans Albert. She wrote to a friend, "I only hope and wish that fame does not have a harmful effect on his humanity." At the same time, she became worried about the attention he received from other women, feeling jealous, angry, and neglected. She became especially annoyed when his girlfriend Marie from 10 years earlier wrote him a congratulatory letter on his appointment to the Zurich professorship—so much so that she wrote to the woman's husband, questioning his wife's intention in sending the letter, causing Einstein embarrassment. But Einstein clearly enjoyed the company of women, both young and elderly, and they easily succumbed to his charms and dapper good looks.

At the end of the summer, before Albert gave a lecture in Salzburg, Austria, Mileva persuaded him that they should take a vacation together in the southern Alps. The excursion appeared to mend some romantic fences: by the end of October, shortly after the Einsteins had moved into a small apartment in Zurich and Albert assumed his teaching duties at the university, Mileva was pregnant again. As her pregnancy advanced, Einstein became engrossed in teaching a course in mechanics and another course in thermodynamics, and he presided over a physics seminar at the university as well.

Einstein's renown continued to grow throughout Europe as he published his ideas and review articles, taught his courses, and delivered lectures to scientific audiences. During his lecture in Salzburg, the curious European physicists were finally able to get a good look at him, and Einstein in turn was able to meet many of his correspondents and critics face to face. He was no longer interested in lecturing on relativity, however, and concentrated his attentions on radiation theory. The lecture he gave on the subject in Salzburg was judged as one of the turning points in the evolution of theoretical physics.

Einstein was nominated for the Nobel Prize in physics in 1910 for his contributions to relativity, but the award was not yet to be his. The Nobel Committee felt Einstein's theory should be experimentally confirmed before he received this honor. After that, the committee nominated him for the prize six more times in 10 years (1912–1914 and 1916–1918) until he finally received it for 1921. In the earlier years, he was nominated not only for relativity theory but also for his statistical work on Brownian motion. Even though general relativity was experimentally confirmed in 1919, there were still some skeptics. The 1921 prize was therefore given for his work on the photoelectric effect, which he had also completed and published during his miracle year.

Not surprisingly, other institutions wanted to have Einstein on their faculties, too. In the spring of 1910, officials at Charles University, the German university located in Prague (considered one of the oldest and most famous institutions of higher learning in Europe), were anxious to bring him to the Czech city. They recommended him for a professorship in theoretical physics, a position that first needed to be approved by the Ministry of Education of the Austro-Hungarian Empire, of which Prague was a part. Einstein was flattered to be considered for the prestigious post and was definitely interested in it, but confirmation would be a long and uncertain process. He felt that part of the problem was his Jewish roots. Anti-Semitism was widespread in the empire and in much of Europe in general at the time.

Because the proposed position would be a promotion over his current professorship, he let it be known in Zurich that he would leave if the offer came through. His enthusiastic and loyal students, hoping to dissuade him from accepting, petitioned the officials in Zurich to encourage him to stay and to offer him a raise. Even though Einstein was not always an effective teacher or lecturer, his students enjoyed his down-to-earth congeniality and his creative ideas and wanted to reap these benefits for themselves. The university officials were persuaded and offered Einstein a raise of 1,000 Swiss francs per year. But Einstein, anxious to improve his standing in the European scientific community so that his ideas could be more widely disseminated, was not convinced that this gesture was enough. He decided to go to Vienna, the seat of the Austro-Hungarian Empire, in early fall to discuss the prospective appointment further with the authorities there. He also met and paid his respects to Viennese resident and philosopher Ernst Mach, whose work in "natural philosophy" Michele Besso had introduced to him and who had influenced his scientific thinking so much in his earlier years. Late in the year, after his return from Vienna and while waiting to hear about the Prague appointment, Einstein published a paper that, among other things, explained the blue color of the sky. It is one of his most difficult papers to understand.

The year had already been full of family milestones. Einstein's sister, Maja, had married Paul Winteler in March; Mileva's parents had come to visit in the summer; and toward the end of July, the Einsteins' second son, Eduard ("Tete" or Teddy), was born. Such events, though they gave him joy, were now taking second place in Einstein's life behind his scientific and professional pursuits, causing Mileva to complain, "With that kind of fame he does not have much time left for his wife." Tete, a sensitive and bright child, turned out to be difficult to raise. Frail and sickly throughout much of his early childhood, he had a few years of good health before he was diagnosed as mentally ill around the age of 20.

In early January 1911, Einstein received some good news. Emperor Franz Joseph had approved his appointment to the chair of theoretical physics at Charles University in Prague, effective in April. In addition, as a bonus Einstein would be appointed director of the Institute of Theoretical Physics. His duties would also include teaching courses in mechanics and thermodynamics that year and leading a physics seminar. The emperor offered him an attractive salary, too—about twice the amount the Swiss could afford to pay. He didn't have to think about his options too long—the opportunity was too good to turn down at this stage of his career. Later in the month, with some personal regret, he sent his letter of resignation to the Zurich officials. At the end of March, after a stopover in Munich, Professor Einstein moved his family east.

Prague, an architecturally charming old city founded in the ninth century, had the reputation of being home to a snobbish and pompous society. Considered a leading academic center of central Europe, this ancient capital developed into a captivating but complex city, with the Czech and German cultures competing with each other and fighting for position. The Germans, making up only 10 percent of the total population, looked down their noses on the other residents, claiming the highest social standing for themselves. The Jewish residents were quite low in social rank. Not long after his arrival, Einstein wrote to Michele Besso that the German residents were "cold and an odd mixture of snobbery and servility, without any kind of benevolence toward their fellow humans." Yet he was greeted as a celebrity as soon as he arrived, with much fanfare and the promise of an elegant social life. Even though he found his colleagues for the most part scientifically unstimulating, he appreciated the beautiful surroundings, the magnificent university library, and the chance to study on his own without major interruptions.

With the generous salary he received, Albert and Mileva were able to maintain a lifestyle that was more bourgeois than the casual one they had enjoyed in Bern and Zurich. They set up their household in a spacious apartment, with enough room for a maid and, for a while, even for Mileva's mother. Amid the new amenities, Mileva could not find happiness, feeling uncomfortable and intimidated among the self-important residents of the town. She tried to find solace in caring for Hans Albert and baby Eduard but became seriously depressed because she didn't fit in and Albert was absorbed by his professional affairs. Even with the Einsteins' affluent existence—or probably because of it—the mood in the household became increasingly gloomy, exacerbated by Mileva's growing jealousy of Albert's fame.

To add to the gloom, Einstein now came down with serious stomach problems that would continue to plague him throughout his life. Though often confined to bed, he remained well enough to fulfill his duties at work. Partly to escape the joyless mood at home and for the intellectual satisfaction he couldn't find at work, he became a regular visitor to one of the fashionable salons of the old city, run by Berta and Otto Fanta. In the Fantas' salon, an eighteenth-century house with an impressive view of the Castle of Prague, young Jewish intellectuals and other notables gathered weekly to chat about philosophy, music, and literature, much as Einstein had done in earlier years with his friends in Bern. Among the salon guests who would read from his manuscripts was Franz Kafka, a sullen young lawyer whose existentialist writings had not yet become known and who was trying to solidify his Jewish identity. Also present were other writers, philosophers, and Zionists. Johanna Fantova, related to the Fantas through marriage, organized Einstein's personal library in Berlin in 1929–1930 and became an intimate friend of his 40 years later when both lived in Princeton.

While he was in Prague, Einstein, still young at the age of 32 but already famous, continued to receive other job offers. The temptation to leave Prague and its social trappings was big, especially on Mileva's part. One of the overtures to steal him away came from Einstein's alma mater in Zurich, the ETH. Because the prospect of returning to the Swiss city was attractive to both him and Mileva, Einstein soon took part in negotiations to secure the position for the following year. Among those giving Einstein a generous recommendation for the position were Marie Curie and the well-known French mathematician and philosopher of science Henri Poincaré, who had also contributed to relativity theory. Einstein spent several months waiting to hear some news from Zurich. In the meantime, he formulated his first decisive ideas on general relativity—the effect of gravity on light—and suggested his theory be tested at the time of the next total solar eclipse.

Toward the end of the year, in early November, Einstein lectured at the first Solvay Congress in Brussels. Established by the benefactor and chemist Ernest Solvay of Belgium, this conference was a grand affair. All the physics luminaries of Europe were present. Einstein was the youngest participant, but he mingled with the others with assurance, grace, and easy humor. Here he again met the widowed Marie Curie, who had just won her second Nobel Prize. According to her, after speaking with Einstein, she was able to "appreciate the clarity of his mind, the breadth of his information, and the profundity of his knowledge. ... One has every right to build the greatest hopes on him and to see in him one of the leading theoreticians of the future." Einstein, on the other hand, although he admired her greatly for her contributions, later described her as being "as cold as a herring. And her daughter is even worse, like a grenadier."

The congress, which assembled to discuss progress in radiation and quantum theory, was such a resounding success that the enthusiastic Ernest Solvay announced that these summits would be held every two years. Now known as the Solvay Conferences, they have become an important assembly point for the world's preeminent physicists and chemists. At their meetings, experts in a given field discuss one or several related problems of fundamental importance and look for ways to solve them.

In late January 1912, Einstein's appointment as professor of theoretical physics at the ETH was confirmed. The family happily planned to leave Prague in July and looked forward to the move back to Zurich. But before their departure, Einstein traveled by himself to Berlin, where members of the extended Einstein family were living. While he paid his respects to his aunt and uncle, he also reacquainted himself with his cousin Elsa, whom he had remembered from family visits when he was a child. Elsa was three years older than Albert, and she was now divorced (her ex-husband had died since the divorce) and had two daughters: Ilse, who was 13, and Margot, 11. Elsa's and Albert's mothers were sisters, and their fathers were first cousins, so the cousins had a complicated family connection. Elsa, with her light blue eyes and sunny and outgoing disposition, was the total opposite of the mirthless Mileva, and Einstein was attracted to the vivacious and energetic woman at once. He came back to Prague with Elsa on his mind and immediately began a secret correspondence with her that he kept up even after he and Mileva moved back to Zurich.

After the return to Switzerland, Mileva wrote to friends that she was happy to be back in Zurich and away from the unhygienic and sooty conditions of Prague that had had adverse effects on her children's health. With Albert's high salary in his new position, the family was able to continue to live as comfortably as in Prague, and they occupied a large, sunny apartment with modern conveniences close to the university. However, regarding her husband, Mileva wrote to her friends, "I must confess with a bit of shame that we are unimportant to him and take second place." While Einstein may have enjoyed returning to the tight circle of his Zurich friends and his teaching duties, Mileva appeared to be in an emotional downward spiral, becoming more and more downcast and morose. Contributing to her anxiety was not only the constant pain in her legs but also the woeful situation in the Balkan states, which were mobilizing for war, causing her concern for family and friends. Einstein, however, could not summon up much sympathy for his wife and increasingly distanced himself from her, often escaping to play the violin in small ensembles with friends.

In 1912, as a contribution to a handbook on radiology, Einstein began to prepare a manuscript on a review of his work on what later came to be called the "special theory" of relativity (to distinguish it from the 1915–1916 "general theory"). It took him two years to finish the work, but then World War I intervened and interrupted its publication. The editor asked him to revise and update the work several years later, but by then he was too busy, and the manuscript remained unpublished for more than 80 years. It was finally published in 1995 in volume 4 of *The Collected Papers of Albert Einstein* and gives valuable insights into Einstein's ideas on relativity up until this time.

Early that year, Einstein had begun one of the closest friendships of his life with the Dutch physicist Paul Ehrenfest, who had visited him in Prague. "Within a few hours we were friends, as if made for one another by our strivings and longings. We remained linked in sincere friendship until he departed this life," Einstein recalled later. They began an enthusiastic, long, and warm correspondence, discussing physics, music, Zionism, and their families. Einstein even procured violins and a piano for the family through his contacts in Germany. For many years, Ehrenfest had tried to lure Einstein to a permanent position in Leyden, but by this time Einstein was already firmly ensconced in his position in Berlin. In 1934, Ehrenfest, suffering from what Einstein called a "morbid lack of self-confidence," committed filicide, then suicide: he shot his 16-year-old son, Vassik, who had Down syndrome, in the hospital where he was being cared for and then shot himself. His death was a profound loss not only for Ehrenfest's wife, other son, and two daughters but also for Einstein.

The Einstein marriage continued to deteriorate in 1913, but for the time being, Albert felt he could not escape from his responsibilities to his wife and young sons. He wrote conflicted letters to Elsa in Berlin, longing for her but feeling trapped. The dreadful pain in Mileva's legs weakened her and made it difficult to get around and attend to the boys and house. To Einstein, she became a tiring and perhaps tiresome spouse, often depressed and complaining—though probably for good reason. Not to be distracted, he kept busy with work on his new gravitational theory and tried to avoid thinking about Elsa. He had stopped corresponding with her nine months earlier for fear of their family's disapproval but resumed writing her after she had sent him a birthday letter. Shortly thereafter, Einstein and Mileva took a trip together to Paris, where Albert gave a lecture and the couple were houseguests of Marie Curie. In late summer, they took nine-year-old Hans Albert with them on a hiking expedition in the Engadine Mountains with Curie and her daughters, who were around the same age. Mileva was unable to hike but enjoyed the fresh air and scenery. Soon after, Paul Ehrenfest and his wife, Tatiana, came to Zurich for a two-week visit, and the two physicists were able to cement their friendship further.

The quantum view of nature received another boost in 1913 when Danish physicist Niels Bohr came up with a new interpretation of the atom. He theorized that electrons traveling around the nucleus of an atom would move only in certain definite, or quantized, orbits. According to this theory, a specific quantum of light, the photon, could be emitted or absorbed when an electron moved from one allowed orbit to another. Bohr won the Nobel Prize in physics for this work in 1922, one year after Einstein would receive his.

In September, the Einsteins took the boys to see their grandparents in Hungary, where Hans Albert and Eduard were baptized in the local Greek Orthodox church at Mileva's request. Their father, who had no interest in organized religion of any kind, did not attend the baptism. After this excursion, they went on to Vienna, where Albert gave a talk on his new gravitational theory, which popularized it and made its way into the mainstream newspapers. Mileva returned home to Zurich, while Albert made a trip to Berlin, where, besides attending to some business, he also met secretly with Elsa.

The same year, Einstein published a paper on general relativity theory, with his friend, mathematician Marcel Grossmann, as coauthor. Einstein explained the physical principles, while Grossmann wrote the section on the advanced mathematics necessary to formulate the theory. The new methods were then tested in another paper with another colleague, and Einstein felt confident they had made a great breakthrough in explaining general relativity—so sure, in fact, that he thought experimental verification was unnecessary. But he was willing to wait for the next eclipse, which he was certain would allow experimentalists to prove his theory.

In November 1913, Einstein was informed of the results of negotiations he had been aware of for several months. He was elected to the Prussian Academy of Sciences and was offered a research professorship at the University of Berlin and the directorate of the soon-to-be-established Kaiser Wilhelm Institute of Physics there, all topped off with a generous salary. Without regard to Mileva's wishes, Einstein accepted the offer. One might wonder why he would want to move back to Germany after professing such hatred for it, especially for the Prussians and their educational system. One of the greatest appeals of the offer was that he was under no obligation to teach. Though he enjoyed lecturing, he felt that teaching took too much of his time away from research, which was always his first love, and now he wanted to devote himself more completely to finalizing his work on general relativity. But no doubt, affairs of the heart also played a role in his decision: Berlin, after all, was where Elsa lived. His decision to accept the offer would profoundly change the course of his life.

Chapter 6

THE EARLY BERLIN YEARS: WAR AND PACIFISM

> Heroism on command, senseless violence, and all the loathsome nonsense that goes on in the name of patriotism—how passionately I hate them!
>
> —From "What I Believe," 1930

Einstein's early "Berlin years" were a time during which his fame increased immensely worldwide. People were now beginning to approach him for his opinions not only on physics but on nonscientific subjects as well. Many of Einstein's nonscientific involvements at this period in his life centered on his pacifist activities and his closer identification with his Jewish heritage.

Before the family had moved to Berlin in April 1914, Einstein had been continuing his relationship with Elsa by mail, and now, with his presence in Berlin, the liaison became even more involved. Though Mileva had never confronted Albert about the situation, her correspondence and conversations with friends show that she was suspicious and understandably unhappy, as he ignored and neglected her and left their apartment in the Wilmersdorf section of the city at will. She had not wanted to move to Berlin in any case, being afraid of Albert's family, who still resented her, and afraid of what lay ahead for her in an unknown environment and culture. It was the second time her husband had taken her away from the security she felt in Switzerland.

On June 28, 1914, at Sarajevo, the capital of Bosnia-Herzogovina, a young Bosnian Slav nationalist, Gavrilo Princip, assassinated Archduke Franz Ferdinand, heir to the Austro-Hungarian Empire, which had annexed Bosnia-Herzogovina in 1908. Austria blamed Serbia for the murder. Serbia, which had designs on Bosnia-Herzogovina, appealed to Russia for help, and Germany mobilized itself as an ally of Austro-Hungary. This led to mutual declarations of war between the great powers across Europe, with Germany's attack on neutral Belgium bringing Britain into the World War I on August 4. Too late, Kaiser Wilhelm tried to scale back German involvement in the conflict: by now, the German military wanted to strut its stuff. The "Great War" had started—the "war to end all wars."

In the midst of this turmoil, Einstein decided to end his own domestic war, drawing up and presenting to his wife in mid-July an extraordinary memorandum outlining under what conditions he would continue to live with her:

> (A) You will see to it that (1) my clothes and laundry are kept in good order; (2) I will be served three meals regularly *in my room;* (3) my bedroom and study are kept tidy, and especially that my desk is left for *my use only*.
>
> (B) You will relinquish all personal relations with me insofar as they are not completely necessary for social reasons. Particularly, you will forgo my (1) staying at home with you; (2) going out or traveling with you.
>
> (C) You will obey the following points in your relations with me: (1) you will not expect any tenderness from me, nor will you offer any suggestions to me; (2) you will stop talking to me about something if I request it; (3) you will leave my bedroom or study without argument if I request it.
>
> (D) You will undertake not to belittle me in front of our children, either through words or behavior.

The French essayist Montaigne once wrote, "There is no man so good that, if he placed all his actions and thoughts under the scrutiny of the laws, he would not deserve hanging ten times in his life—yes, even the kind of man whom it would be a great scandal to punish and a great injustice to execute." Reading Einstein's memo, one is tempted to pass sentence on him, now a frustrated and unhappy man in his personal life. He and Mileva had been married 11 years, many of them unhappily. In letters to Elsa in December 1913, he had written, "Do you think it's so easy to get a divorce when one has no proof of the other party's guilt? ... I am treating my wife like an employee whom I can't fire. ... She is an unfriendly, humorless creature who gets nothing out of life and who, by her mere presence, extinguishes other people's joy of living." The memo to Mileva was no doubt deliberately designed to humiliate and alienate her further, spelling out in cold, cruel words how he intended to continue treating her—at least this is how he felt now, in the heat of his anger toward her. He wanted her to get the message about their relationship and depart, and apparently he couldn't find in himself a kinder way of doing so.

At first, Mileva, so dependent on her husband financially and socially, accepted the conditions. But the humiliation and emotional turmoil must have been too great, for in late July she packed her bags and abruptly left Berlin with the boys, returning to Zurich. She prepared herself and her sons for an eventual divorce. This new freedom now made it easier for Albert to continue his liaison with Elsa. He wrote to his friend Besso, telling him he was content with the separation, which brought him peace and quiet at home and also made possible "the excellent and truly enjoyable relationship with my cousin. Its stability will be guaranteed by the avoidance of matrimony." By this time, Einstein had arrived at a bitter view of marriage, whose attendant trappings he feared made his lifestyle too bourgeois. "Marriage makes people treat each other as articles of property and no longer as free human beings," he claimed. And "marriage is but slavery made to appear civilized."

Now, as war was breaking out all over Europe, Germans at all levels of society were expected to support the war effort in some way. Being a Swiss citizen, Einstein was not obligated to do so. Indeed, he declared himself a pacifist and, for the first time, began to speak publicly about his feelings on war as those around him, caught up in patriotic fervor, cheered at the early German military victories. "Europe in its madness has now embarked on something incredibly preposterous," Einstein wrote to his friend Ehrenfest in August 1914. "At times such as these one sees what deplorable beasts we are. ... I can feel only a mixture of pity and disgust." He signed a pacifist manifesto, drafted by a family friend and physician, Georg Nicolai, that appealed to European rather than nationalist German sympathy and unity. He also attended antiwar meetings, one of them held by a pacifist group devoted to establishing a United States of Europe, which the government subsequently banned in 1916. None of Einstein's appeals and efforts, however, made any difference.

Einstein's pacifism was visceral. Ever since childhood, he had detested the military. He could not understand the willingness of soldiers to be obedient to their superiors and to their governments without thinking independently about the rights and wrongs of war and policy. "My pacifism is an instinctive feeling, a feeling that possesses me because the murder of people is revolting. My attitude is not derived from any intellectual theory but is based on my deepest antipathy to every kind of cruelty and hatred," he would later write to the editor of a Christian periodical. He felt the most important function of a government is to protect individuals and allow them to develop into creative persons, and forcing them to serve in the military violates this principle. Even as a young man of 22, he had written to "Papa" Winteler, "A foolish faith in authority is the worst enemy of truth." He disdained any kind of competition where one person or group tries to outdo another, even in harmless and creative games such as chess. He could not understand people's fondness for competition and the pleasure they received from winning, not only in war but in any kind of struggle for power and status.

Einstein, in his new position, had to be tactful with his German colleagues and benefactors, many of whom he privately scorned for their ideology. Mindful of his status as a noncitizen, he did not argue or confront them but rather asked them questions to stimulate alternative thinking. He thought of himself as a citizen of the world, not of any particular country. "Nationalism is an infantile disease. It is the measles of mankind," he remarked. Part of this internationalist feeling came from his feeling that scientific research should be shared freely throughout the world and that pitting one nation against another makes it difficult for scientists to engage in professional exchanges.

Curiously, he did not seem particularly perturbed when some of his friends contributed to the war effort through their expertise in explosives and chemicals. Einstein's friend, chemist Fritz Haber, was put in charge of directing Germany's chemical warfare activities and initiated the use of poison gas. Another Einstein colleague, Walther Nernst, a board member and trustee of the Kaiser Wilhelm Institute, also became a leader in this effort; both of Nernst's sons were subsequently killed during the war. Haber won the Nobel Prize in chemistry in 1918 and Nernst in 1920.

As World War I continued to plague Europe and 10 years after publishing his first paper on relativity theory, Einstein struggled intensively to extend his work on relativity. At this time, many physicists continued to be skeptical about the theory because they did not have the necessary tools and equations to understand it, and no one had yet proven it experimentally. Others just ignored it. Nevertheless, Einstein steadfastly continued to clarify his ideas in his publications and lectures, managing to produce 17 talks and publications that year. Throughout World War I, he worked on what later came to be known as "general" relativity.

The year 1914 was also one in which a solar eclipse was to appear, and the best place to see it was in southwestern Russia. A German colleague of Einstein's, Erwin Freundlich, had made arrangements to go to Russia in August to take a series of photographs of the eclipse that would prove one of Einstein's predictions of relativity theory—that light would deflect, or bend, in a gravitational field. Shortly after Freundlich and his group arrived in the Crimea, the war started, and Germany and Russia became instant enemies. The scientists were captured and placed in internment camps, and their equipment, borrowed from an Argentine expedition, was confiscated. As a result, neither the Germans nor the Argentines could take photographs—but neither could the members of an American expedition, who were not enemies of Russia because of the early neutrality of the United States in the conflict. Bad luck was on their side, for thick gray clouds obscured the eclipse on August 21. The next opportunity to prove Einstein's theory would not come until 1919.

By the end of 1915, Einstein had a breakthrough and came up with a completed version of his generalized theory of gravitation (see chapter 7). He delivered four lectures to the Prussian Academy of Sciences summarizing his exciting results. His 1915 theory replaced the Kepler-Newton theory of planetary motion, which was based on the assumption of absolute space. The new theory was able to account for the slow rotation of the orbital ellipse of a planet. Einstein was now able to predict radically new phenomena—the bending of light around the Sun and the precession of the perihelion of Mercury.

In the meantime, Einstein's intensive work and his continuing affair with Elsa were blinding him to the responsibilities of fatherhood. His sons, particularly Hans Albert, were beginning to resent him and his often-unreasonable demands of them when he did write. Though he missed his children and worried about them, he knew they were safer in neutral Switzerland than in Germany. He wanted to visit them in Zurich in the summer, but Mileva said they would be away, so he vacationed on the Baltic Sea with Elsa and her daughters instead. He kept up his correspondence with the boys and Mileva, however, and visited them in September, taking Hans Albert on a hiking trip in the mountains.

Einstein made his feelings about the war known in a short essay, "My Opinion on the War." It was written in October–November 1915 at the request of the Goethe-Bund, a German cultural organization, for a volume of "patriotic commemoration" to be published the following year. In it, well-known Germans were called on to defend German culture in the midst of war. Instead of making a patriotic declaration, however, Einstein wrote about his distaste of war. He stated that war is rooted in the "biologically determined aggressive tendencies of the male," and he upheld pacifism and rejected war under any circumstances. Much of his original statement was not published. The following words were censored: "The state to which I belong as a citizen plays not the slightest role in my emotional life. I regard a person's relationship with the state as a business matter, akin to one's relationship with a life insurance company." According to pacifist Romain Rolland, whom Einstein had met in September in Switzerland when he went hiking with Albert, Einstein hoped the Allies would win the war so that the power of the Prussians would be quashed. He favored splitting Germany into two parts based on geography, with Prussia in the north and southern Germany and Austria in the south.

In 1916, Einstein became president of the German Physical Society, a position he held until 1918. But more important, his long-awaited details on general relativity were published. The title of this seminal paper was "Foundations of the General Theory of Relativity," and from then on, the 1905 theory became known as "special relativity" or the "special theory" and the 1915–1916 theory, which included the phenomenon of gravitation, as "general relativity" or the "general theory." The general theory proposed to explain all laws of physics in terms of mathematical equations (see chapter 7). Because many people found the mathematics quite complicated, Einstein wrote a new, less technical exposition and published it the following year as *On the Special and the General Theory of Relativity, Generally Comprehensible*, which became his best-known book. Despite its title, it remained incomprehensible to many readers. (An English translation appeared in 1920, titled *Relativity, the Special and General Theory: A Popular Exposition*.) Perhaps more comprehensible were his contributions to quantum mechanics in 1916, which resulted in three papers.

Einstein continued his pacifist activities during a year in which millions were killed all over Europe, both in the east and in the west and in the seas. Germans were still in the wake of the nationalistic wave set in motion by Chancellor Otto von Bismarck some 40 years earlier when he unified the German states and would not end until it had run its course after World War II. When Kaiser Wilhelm II came to power in 1888, he too had expansionist views, but he dismissed the elderly, more politically savvy Bismarck within two years. The German people now had a leader who was trying to repeat the conquests of the past while lacking the skills to do so.

Most Germans, both progressives and conservatives, were in support of the war at the start. The belief was widespread that even though their leaders had forced the war on them, it was for their common good. Further into the war, however, most of them were willing to negotiate a peace as long as they did not have to sacrifice any land, but the government did not want to be seen as a loser and stuck to its plan, also fearing a revolution in favor of more liberal policies. As the war continued, the progressives split up into different factions, many of them opposing any kind of war and thereby revealing the first seeds of discontent among the people. Importantly, the German economy was suffering as well. The government did not have large stockpiles of guns or ammunition and had to employ chemists, such as Einstein's friend Fritz Haber, to find replacements for substances made scarce by the war. The shortage of goods in general, much of it due to the palpable labor shortage after the male population was drafted, led to riots in 1916.

Einstein managed to send letters to his sons in neutral Switzerland even during wartime, telling them he would visit them as soon as peace came to Europe again. He tried to improve his relationship with Hans Albert, telling him, among other things, not to worry about his schoolwork too much: "Don't worry about your marks. Just make sure you keep up with your work and that you don't have to repeat a year. But it's not necessary to have good marks in everything." He also advised that on the piano, he should play only the things he enjoys, even if the teacher didn't assign them to him. "You learn the most from things that you enjoy doing so much that you don't even notice time is passing. Often I'm so engrossed in my work that I forget to eat lunch." But the father was not so easygoing about spelling errors: "You still make so many writing errors. You must take care in that regard: it makes a bad impression when words are misspelled." Finally, he told his sons, they should not forget to brush their teeth every day.

Early in 1917, Einstein displayed symptoms of serious gastrointestinal problems, and within two months he lost 56 pounds. It would be four years before he overcame this affliction. Elsa was only too happy to take care of him since he was now conveniently situated right across from her fourth-floor apartment in the Schöneberg district of Berlin. By December, he was ill again with an abdominal ulcer and was confined to bed for several months. Einstein, too weak from his stomach problems even to submit his own papers to the various physics journals, had to enlist his colleagues to do so for him.

Meanwhile, Mileva's health problems continued to increase in 1917 as well as she became depressed and bedridden and was confined to a sanatorium. Luckily, Einstein's friends Heinrich Zangger and Michele Besso were willing to look after the boys. They blamed Mileva's breakdown on Einstein's treatment of her, and Einstein accepted partial responsibility for her condition. Mileva, he learned, was still hoping for reconciliation, but Einstein, now in love with Elsa, ruled it out. He realized it was unlikely that Mileva would agree to an amicable divorce right now, so he dropped the subject for the time being. For now, Einstein wondered whether he should bring at least Hans Albert to Berlin but was afraid of Mileva's response to the suggestion. He was sending more than half his salary to Mileva and the boys and additional funds to his mother, who had become ill with stomach cancer. He worried about his ability to take care of everyone properly since the German mark was so deflated and didn't go far in Switzerland. Though many people in Germany were literally starving and good food was scarce, he was able to get produce packages from relatives in the rural south and dairy products from Heinrich Zangger in Switzerland. Because of his digestive problems, it was especially important for him to adhere to a strict diet with the proper foods.

In October 1917, while Germany was still entrenched in the war and about three and a half years after Einstein's arrival in Berlin, the Kaiser Wilhelm Institute of Physics finally opened, with Einstein as its first director. Its mission was to promote research in physics and astronomy. A few months later, Einstein was feeling better and was able to get out of bed to attend a few professional meetings and, not long afterward, to teach some courses at the University of Berlin. Though he was not obligated to teach, he could do so at his pleasure.

In 1917, the Germans had begun to sink American ships that were carrying munitions and reinforcements to the Allies, thereby drawing the United States into the war. The Allied armies were greatly helped by American involvement in the war, which rose to over 2 million soldiers. Over the next year, Germans became increasingly fed up with the war and pressured their government to end it. In 1918, after they had suffered huge losses in the west after gaining earlier victories in the east, German officials realized the end was near. They began to negotiate peace treaties with various nations and finally proposed an armistice. The Allies insisted they would grant it only if Germany made no further move to improve its military strength and mandated that the Germans retreat into their own borders and leave all their weapons behind. Germany reluctantly accepted the terms, and World War I ended in November 1918 with the abdication of Wilhelm II in Berlin. The war dead on both sides of the conflict numbered in the millions.

The toll of war was terrible. *The New York Times Almanac* provides a useful summary: "Over eight million died in battle and six million civilians perished." As a consequence, "the war turned over the old European state system as four empires collapsed and were partitioned." The four empires were the German, Russian, Ottoman, and Austro-Hungarian. Out of the ruins would emerge Nazi Germany and Soviet Russia and their dreams of world domination. As if Nature were mocking the destructiveness of humankind, in the last months of the war, an influenza epidemic began to sweep the world, taking between 20 and 40 million lives overall before it ended in 1919. It was the worst epidemic in terms of its virulence the world had ever known, even worse than the plague in medieval Europe. The death toll in the United States was about 675,000 and in Europe 2.3 million. The airborne virus, an insidious enemy attacking soldiers and citizens on both sides of the conflict as well as people in countries that were not at war, hit the closely quartered soldiers hard: more American soldiers, for example, died from the pneumonia caused by the flu than died in combat. Because the incubation period of the virus was only a couple of days, quarantining the victims did not help much, and antibiotics did not yet exist.

All the Einsteins, even those who were weakened by illness, survived both war and disease. But even after the war, the times were hard. Food was still scarce in Europe, and many Germans and Austrians were starving. Humanitarian and faith-based groups such as the Quakers busied themselves in providing relief for children. Einstein, like many others, was distressed by the high child mortality rate and the mass hunger, especially as he saw it all around him in Berlin. He publicly thanked American and British Quakers for their efforts at providing food for more than half a million German children, a task in which the newly formed League of Nations involved itself, too.

Einstein now became optimistic that a free and democratic society would be possible in Germany. He committed himself to the democratic-socialist goals that became popular among intellectuals in Europe at the time. Although Einstein was not a socialist by today's definition, his politics were left of center, much like today's liberal wing of the Democratic Party in the United States. Because he was not personally interested in material success and in accumulating wealth and because he spoke out against the excesses of capitalism, people have generally thought of him as a socialist. But there is no evidence he objected to the ownership of private property—he would own his own home in about 10 years—or running a private business as his family had always done. There is no doubt, however, that he favored greater government involvement in social services, education, and health care, as is common even in nonsocialist Europe. A believer in freedom of speech and opinion, he was not afraid of being seen with Communist sympathizers when their goals seemed justified to him, but he was not interested in advocating Communist-dominated causes.

The Einstein-Marić marriage had no chance of surviving, and Einstein had already signed a divorce agreement with Mileva during the summer before the war ended. According to its terms, Einstein would deposit a substantial amount of money into a Swiss bank account for Mileva's use and also make quarterly alimony payments. Furthermore, if he were to receive the Nobel Prize, this money would also go into the account. Mileva could spend the interest from it as she pleased, but she needed approval from Albert to use any of the principal. Mileva would have custody of the children, but Einstein was free to spend time with them whenever he was in Switzerland during school holidays. Toward the end of 1918, as the war ended, the couple took the first legal steps to end their marriage formally. First, Einstein had to give a reason for wanting to end the marriage, so he formally declared himself an adulterer and paid a fine and court costs for the offense. Second, the court forbade him to remarry for at least two years, at least in Switzerland. Just before the divorce was final, Einstein went to Zurich for a month to deliver the first set of lectures as a visitor at the University of Zurich and to spend time with Mileva and his sons, all of whom had, by this time, reconciled themselves to their fate.

While Einstein was in Zurich, Berlin found itself in chaos again. The ill-organized, extreme left-wing "Spartacist uprising" organized by the German Communist Party took place in early January. The Spartacist League, named after Spartacus, the leader of the largest slave rebellion in classical times, tried to gain control of Berlin. The attempt was crushed violently by the combined forces of the German Social-Democratic Party, the remnants of the German army, and the right-wing paramilitary group known as the Freikorps. At this time, the Freikorps consisted mostly of soldiers who had returned defeated from the war and favored a return to a military structure in government; others joined up in an effort to ward off Communism. Rosa Luxembourg and Karl Liebknecht, the league's leaders, were captured and murdered, along with many others, while held prisoner by the Freikorps, and their bodies were dumped in a river on January 15.

The year following the end of war was a watershed in Einstein's personal life: he divorced Mileva; he married Elsa, who would remain his wife until her death 17 years later, despite his misgivings about marriage; and he became an international celebrity after his theory of general relativity was confirmed (see chapter 7). Besides devoting himself increasingly to problems of international reconciliation after the war, he also began to identify increasingly with the Jewish people. Many of them were heading west from eastern Europe, where they had long been facing discrimination, seeking a better life in Germany and other Western nations. But things weren't much better in Germany, where they found a similar bias, even among many German Jews who considered themselves more culturally and socially advanced than their mostly agrarian kinfolk from the east.

Through his friendship with Zionist activist Kurt Blumenfeld, whom he later thanked for having "helped me become aware of my Jewish soul," Einstein became interested in Zionism and began to support the idea of a Jewish state in Palestine. Until he had come to Berlin, his Jewishness had never been an issue with him, for he was completely neutral about ethnic distinctions and felt at home anywhere he went, among diverse groups of people. He supported the Zionist ideals of a Jewish homeland, but he never joined any Zionist organizations. He was opposed to nationalism of any kind, even Jewish nationalism, which Zionism represented. But he supported the creation of Israel as a refuge for Jews because he believed in the power of community as a cohesive cultural force. In 1920, he told a German Jewish organization, "I am not a German citizen, nor is there anything in me that can be described as a 'Jewish faith.' However, I am happy to be a member of the Jewish people, even though I do not regard them as the Chosen People." He expressed the belief that anti-Semitism itself may have preserved the Jewish race; that is, the continuous fight against prejudice kept Jews together in an international fellowship. Without it, he felt, Jews would have been quickly assimilated into the societies in which they were living. He spoke out against limiting Jewish ethnic nationalism to Palestine, believing Jews should be acknowledged in the Diaspora wherever they lived; otherwise, without a visible and vocal presence, the world would deny the existence of a Jewish people. He felt American Jews were more aware of this fact than German Jews, who seemed less concerned about keeping their Jewish identity.

Early in the year, shortly before his divorce from Mileva, Einstein had moved across the hall into Elsa's apartment, where she gave him his own room and study. In February, on Valentine's Day, he and Mileva were formally divorced. Albert and Elsa were thus free to marry in June. The obligatory two-year waiting period set by the Swiss judge was not enforceable in Germany. By now, Ilse was 22 years old and Margot 20, and both young women continued to live with the Einsteins. Also in the Einsteins' Berlin residence lived Elsa's mother, Fanny (who was also Albert's aunt, that is, his mother Pauline's sister), and the sisters' brother, Jacob. Only one year earlier, Ilse, a pretty young woman with one glass eye, had confided to family friend Georg Nicolai, the physician and pacifist mentioned earlier, that Einstein had proposed marriage to her as well—with her mother's knowledge. Ilse had turned him down, saying that her love for him was more like that for a father than a prospective spouse. After the rejection by the younger woman, Einstein appeared to be just as content to marry Elsa, and Ilse became his secretary instead.

Just four days before the wedding took place on June 2, another solar eclipse occurred. A British scientific expedition, led by Sir Arthur Eddington, England's foremost expert on Einstein's theory of general relativity and director of the Cambridge Observatory, had already set off for the island of Principe off the West African coast. His mission was to measure the bending of starlight by the gravitational pull of the Sun. Einstein was confident the team would prove his theory correct.

Young Albert Einstein at age five, with sister Maja, age three, in 1884. (Lotte Jacobi Archive, University of New Hampshire)

Einstein's college girlfriend, Mileva Marić, in 1896. In 1903, she became his wife. (Schweizerishes Landesbibliothek, Bern)

Einstein at around age 21, ca. 1900, while a student in Zurich. (Lotte Jacobi Archives, University of New Hampshire)

Einstein's sons Edward ("Tete"), left, and Hans Albert ("Adu") in Switzerland, ca. 1918. (Leo Baeck Institute)

Einstein, ca. 1922, at age 43, around the time he made his first trip to the United States. (AIP Niels Bohr Library)

With actor Charlie Chaplin at the premiere of the film *City Lights*, January 1931, while Einstein was on a trip to the United States. (Leo Baeck Institute)

With Elsa in Pasadena, 1931. (Emilio Segrè Archives, AIP)

While a professor at the Institute for Advanced Study, Princeton, ca. 1935. (Courtesy Department of Physics, Princeton University)

Sailing at Huntington, Long Island, New York, in 1937. (Lotte Jacobi Archives, University of New Hampshire)

After obtaining his U.S. citizenship, with stepdaughter Margot and the presiding judge, Trenton, New Jersey, 1940. (Courtesy of Todd Yoder)

At Lincoln University, a university for black students about 45 miles southwest of Philadelphia, where Einstein received an honorary degree in 1946. (Leo Baeck Institute)

In his office at the Institute for Advanced Study, late 1940s. (Courtesy of Arts Council of Princeton)

In fuzzy slippers, late 1940s. (Courtesy of Gillett Griffin)

In old age, in his office at the Institute for Advanced Study, early to mid-1950s. (Fantova Albert Einstein Collection, Princeton University Library)

Forever carved, among others, into the archway of the Riverside Church, New York City, the only living person, scientist, and Jew so honored. (Photo by Alice Calaprice)

Chapter 7

THE ROAD TO GENERAL RELATIVITY

After a careful study of the plates, I am prepared to say that there can be no doubt that they confirm Einstein's prediction. A very definite result has been obtained that light is deflected in accordance with Einstein's law of gravitation.

—British Astronomer Royal, Sir Frank Dyson, after the Eddington expedition in May 1919 confirmed Einstein's general theory of relativity

Dear Mother, Today I have some happy news. H. A. Lorentz telegraphed me that the English expeditions have really verified the deflection of light by the sun.

—To Pauline Einstein, September 27, 1919, telling her about the experimental confirmation of the general theory of relativity

The *Annalen der Physik,* in the early twentieth century, was the world's most prestigious physics journal. It was the *Annalen* that published Einstein's key papers of 1905. Every physicist in the German-speaking world would see them. His work was controversial, to say the least, and so Einstein expected, as his sister, Maja, wrote in her short biography of him, "sharp opposition and severest criticism." She went on to report that "he was very disappointed. His publication was followed by an icy silence." The silence would soon be broken. German physicist Max Planck asked Einstein to explain some of the more obscure points in the theory. Einstein did so, and Planck became the first physicist besides Einstein to write an article about relativity. A breakthrough came when one of Einstein's former professors at the ETH, Hermann Minkowski, turned his attention to Einstein's special theory. Minkowski, a mathematician, coined some of the phrases that have become household words among physicists, such as *light cone* and *worldline*. More important, he brought special relativity to the attention of the mathematics community.

* * *

Special relativity used two reference frames, one moving at constant velocity in a straight line away from the other. Einstein thought that most of physics could fit in with his new theory. The one area of physics that posed problems, though, was gravity. What would happen if one of the frames were accelerating relative to the other? Could the answer be the key to incorporating gravity into Einstein's new physics?

Even before special relativity, acceleration had caused problems for physicists. In an accelerating frame, strange "fictitious forces" occur. Suppose you go around a traffic circle in a car traveling counterclockwise, perhaps a bit too fast. You are thrown to the outside of the car, that is, to the right. You feel a force pushing you away from the center of the circle. This is the *centrifugal force*, from two Latin words meaning "flying from the center," but the force is not real. The turning force provided by the car tries to make it move perpendicular to the direction in which you're currently going, but the inertia of the car tries to get it to continue in a straight line. It's the fight between the turning force and the inertial force that gets the car to go safely around the circle. The fictitious centrifugal force is a consequence of the struggle between these two.

To account for acceleration, Einstein needed inspiration. In 1907, he was struck by what he later called "the happiest thought of my life." He had been asked to write an article that summed up the current status of special relativity. The invitation came not from the *Annalen der Physik* but from the famous German physicist Johannes Stark, editor of the *Yearbook of Radioactivity*. Einstein worked hard planning the article, thinking about how he might incorporate gravity into relativity. He later recalled that he "was sitting in a chair in the Patent Office in Bern" when he realized something important: if a man falls off a roof, he cannot feel his own weight. This was the breakthrough he needed. His idea became known as the "principle of equivalence," which states that a reference frame placed in a gravitational field is identical to a reference frame at rest but subject to an acceleration. He included this concept in his review article.

The idea has a long history. In a famous sixteenth-century experiment, possibly only a legend, Italian scientist Galileo Galilei dropped balls of different mass from the top of the leaning tower of Pisa. They landed at the bottom together. This is profound. The gravitational force is proportional to the *gravitational* mass of an object. According to Isaac Newton, a nongravitational force causes an acceleration that is proportional to the *inertial* mass of the object. Galileo's experiment suggests the two are the same, though there is no reason why they should be. Einstein built on this and said, in the equivalence principle, that gravitational and inertial masses are identical. Many years later, in 1964, Missouri-born physicist Robert Dicke would re-create a high-tech equivalent of Galileo's experiment. He placed three masses, two of aluminum, one of gold, on a hypersensitive torsion balance and launched it into space. In weightless conditions, where the orbit of the spacecraft nullifies the Earth's gravity, the entire torsion balance should fall toward the sun. If gravitational and inertial mass are different, the aluminum and gold should fall toward the sun at slightly different rates, which could be measured easily by the torsion balance. Dicke found no significant difference. Einstein's theory had passed a stringent, high-precision test. Bob Dicke, a physicist who held patents on things as diverse as clothes dryers and lasers, deserves a special mention in the history of relativity. Where many physicists succumbed to the elegance, simplicity, and beauty of Einstein's equations, Dicke asked the profound question, Why should nature be elegant, simple, or beautiful? He set about constructing a rival theory that would pass all the same experimental tests that Einstein's did. To show himself wrong and Einstein right, Dicke devised many clever, intricate tests. It is thanks to Dicke that relativity became an experimentally tested theory, one of the most accurate theories in the world of physics.

With the principle of equivalence in his toolbox, Einstein made progress. The 1907 article, which he called "On the Relativity Principle and the Conclusions Drawn from It," arrived on Stark's desk on December 4. It provided a major stepping-stone from the special theory of relativity to the general. Equivalence meant that persons in a closed box could not tell if they were in a gravitational field or whether they were in gravity-free space, accelerated by some unknown force. This situation allowed Einstein to use a mathematical trick. Instead of worrying about the complex details of gravity, all he had to do was work out the mathematics of a constant-acceleration frame. Then, using the equivalence principle, the same equations will hold true for gravity.

Einstein asked what the difference would be between two clocks, with one accelerating relative to the other. After showing that there will be a time difference between the clocks, he transformed this equation from acceleration to gravitation and found that light, in a gravitational field, appears redder. This is the *gravitational redshift*, a major experimental prediction of general relativity. Next, he explored Maxwell's equations, which describe the wave properties of light. Einstein worked out the implications for an accelerating frame and concluded that a gravitational field bends the path of a beam of light. Bending of light is also a major experimental test for Einstein's theory. On Christmas Eve 1907, shortly after sending off his article, Einstein thought of another challenge. He wrote his friend Conrad Habicht, saying he hoped to explain the motion of the perihelion of the planet Mercury. After that, he made no progress on general relativity for almost four years.

The *Annalen der Physik*, under the editorship of Wilhelm Wien, published Einstein's important paper of 1911, "On the Influence of Gravitation on the Propagation of Light," written while he was in Prague. In 1907, Einstein had tried to reconcile Newtonian gravity with special relativity. Now he went one step further. He tried to use the principle of equivalence to construct a new theory of gravity, but he foresaw some problems. He wrote to his colleague Jacob Laub that "the relativistic treatment of gravitation creates serious difficulties." He suspected that one of the building blocks of special relativity—that the speed of light is constant—could hold true only for constant gravity. The 1911 paper covered much of the same ground as the one of 1907. He showed again the gravitational redshift of light and predicted that a light ray grazing the sun would be deflected by a tiny amount, 0.08 arc seconds. Einstein wrote that it "is urgently desirable that the astronomers concern themselves with the questions brought up here, even if … sufficiently unfounded or even adventurous." He did not know that the experiment he asked for, in a few years' time, would bring him international fame.

General relativity was hard going. In 1912, Einstein wrote to physicist Arnold Sommerfeld that "in all my life I have labored not nearly as hard," and then, in a letter to Michele Besso, he complained that "every step is devilishly difficult." Soon, as Einstein finally came to understand he had been using the wrong mathematics, these difficulties melted away. Naturally enough, he had been using Euclidean geometry, the branch of geometry taught in high schools, which deals with flat surfaces, straight lines, and parallel lines that travel to infinity without ever meeting. One cannot describe gravitation that way.

In Euclidean geometry, the shortest path between two points is a straight line. If you open up an atlas and draw a straight line between London and New York City, it will cross mostly over the open blue waters of the Atlantic. Now take a globe and stretch some string between London and the Big Apple. Because the Earth is a sphere, the shortest path between two points is not a straight line but a small arc of a great circle. Airlines know this full well. British Airways planes fly from London, cut across the southern tip of Ireland, brush the tip of Greenland, and fly over Nova Scotia and Boston before touching down at John F. Kennedy Airport in Queens, New York. Airlines use this path because it is the shortest and therefore the quickest and cheapest route to fly. Physicists know that light also takes the shortest path it can. Like the airplane, a photon will take a different route on a curved surface than on a flat one. The shape of a curved surface is determined by what is called its *metric*. Roughly speaking, the metric indicates the distance between two points that are close together on the surface. The shortest path between two distant points on a curved surface is called the *geodesic*. On a flat surface, the geodesic is a straight line. On a sphere, the geodesic is a great circle. Einstein needed to develop some mathematical skills to do with geodesics and metrics that he could apply to nonflat surfaces.

Back in his student days, Einstein had a close friend, the mathematician Marcel Grossmann. For his Ph.D. thesis, Grossmann had discovered some properties of nonflat surfaces. Einstein, on returning from Prague to Zurich, pleaded with Grossmann, "You have to help me or else I'll go crazy." His friend helped Einstein find the mathematics he needed: Riemannian geometry. Einstein wrote to Sommerfeld again: "I have become imbued with a great respect for mathematics, the subtler part of which I had, in my simple-mindedness, regarded as pure luxury until now." Working together, Einstein and Grossmann published papers that introduced Riemannian geometry into the world of physics. In 1913, they derived Maxwell's equations for curved surfaces. A year later, Einstein found an equation that relates the shortest path on a curved surface, the geodesic, to the curvature of the surface. If you know how the surface is curved, then the geodesic determines the orbits of, say, a planet on that surface.

In 1915, Einstein's paper "On the General Theory of Relativity" was published in the *Proceedings of the Prussian Academy of Sciences*. The article pleased him, for it contained a mathematically consistent theory of gravity, one he called "the most valuable discovery I have made in my life." He also said that "the magic of the theory will hardly fail to impose itself on anybody who has truly understood it." The theory's crowning glory is an equation that predicts what the curvature of space is when matter or radiation is present. On the left-hand side is a term that represents the curvature of space; on the right-hand side is a term, the *stress tensor*, which models the effect of matter. Einstein summed up the result this way: "Matter tells space how to bend; space tells matter how to move." The paper presents revised calculations for the bending of light by a gravitational field, which was now twice what Einstein had predicted in 1911.

He also presented the gravitational redshift and had a perfect explanation for the perihelion of the planet Mercury.

Since 1859, astronomers had known something was wrong with the orbit of Mercury. An isolated planet should orbit a star in a perfect ellipse, as the work of Tycho Brahe, Johannes Kepler, and Isaac Newton had already shown. The data did not quite match the predictions, however. An ellipse has a major axis that points in a certain, fixed direction. The planet Mercury, astronomers saw, did not complete an ellipse in one year, and they noticed that the point of its closest approach to the Sun, called the *perihelion*, was shifting. It was like doing a dot-to-dot picture and not connecting on the last dot. The shift in perihelion, its precession, was a mere 43 arc seconds per century as reported in a paper by Erwin Freundlich in 1913, but it was enough to worry astronomers. After 1915, they could stop worrying. The general theory of relativity had explained the precession completely.

Einstein wrote a summary of his new theory, "The Foundations of the General Theory of Relativity," publishing it in the *Annalen der Physik* in 1916. It is a comprehensive description of the final version of the theory. General relativity was now complete. It had predictions that could be tested by experiment, it explained a long-standing problem in astronomy, and—for weak gravitational fields and slowly moving objects—it simplified to Newton's theory of gravity. In the same year, he published the paper "Approximate Integration of the Field Equations of Gravitation," which appeared in October in the *Proceedings of the Prussian Academy of Sciences* and in which Einstein predicted that gravitational waves could exist. The time had come to test the theory and spell out the consequences.

* * *

Curvature and gravity coexist in simple physics experiments, not just in relativity. Think of a droplet of water, usually shaped like a pear, that is about to fall from a faucet. Two forces compete. The gravitational force wants the droplet's center of mass to be as low as possible, so it tries to stretch the droplet vertically to reduce its gravitational energy. The droplet, though, consists of water, and water has a characteristic surface tension. As gravity stretches the droplet, the surface area of the droplet increases. Because of surface tension, the greater the surface area, the greater the surface energy of the droplet. The pear shape of the droplet is a delicate balance between the surface tension force and the gravitational force. Two eighteenth-century scientists, Frenchman Pierre-Simon de Laplace and Englishman Thomas Young, showed that this surface-tension force was related to the curvature of the droplet's surface. So, just as with general relativity, the shape of a droplet is a balance between gravity and curvature.

Another simple analogy with general relativity is a basketball on a trampoline whereby the heavy basketball deforms the trampoline's surface. As the basketball sinks down in the middle, the surface of the trampoline curves gently upward toward the frame. Now, if you throw a marble onto the trampoline, it will start to roll on the trampoline's curved surface. The marble orbits the basketball, just like the Earth orbits the Sun. The basketball tells the trampoline surface how to stretch (matter telling space how to bend), while the trampoline tells the marble how to orbit (space telling matter how to move). If you used a bowling ball instead of a basketball, the heavier bowling ball would curve the trampoline surface more, and the radius of the marble's orbit would be smaller. Similarly, the more massive a star, the greater it curves spacetime around it.

In 1916, German physicist Karl Schwarzschild took a major step by applying Einstein's new theory to astrophysics. He wondered what would happen if you had an isolated blob of matter—a star, perhaps. Because it was isolated, there was no mass in the immediate vicinity, so the stress tensor would be zero. He then assumed that things did not change with time and that there was no preferred direction. Schwarzschild derived the metric for spacetime in the neighborhood of a star. This *Schwarzschild metric*, named in his honor, has a bizarre property: the curvature of spacetime becomes infinite at a certain radius, which is rather like walking down a gentle slope but suddenly stepping off the edge of a cliff—once you've gone too far, there's no way back. This particular distance, called the *Schwarzschild radius*, is the radius of a black hole. Anything that comes within the Schwarzschild radius of a black hole will never be able to escape its clutches.

From Einstein's work on light quanta, we know that energy and mass are interchangeable. It would take a few years before another physicist, Roy Kerr, looked at solutions for an isolated, static, homogeneous region of space that contained not mass but radiation. He found solutions for a rotating, radiating ball of space. Sometimes known as the *Kerr-Schwarzschild metric*, this work helps mathematicians explain the behavior of rotating black holes. It also helps astronomers describe neutron stars, which are highly dense, compressed stars, and their close cousins, pulsars, which are rapidly rotating neutron stars.

Physicists sometimes make tremendous simplifications. We know that, in our local part of the universe, there are planets, dust, stars, and a few galaxies. To make things simple, physicists ignore these niceties and propose that the universe is made up of a nice, smooth "gas" of particles and radiation. For a gas, it's fairly straightforward to work out what the stress tensor should be. With the stress tensor known, the curvature of space can be calculated. The answers you get depend on some other simple assumptions you care to make. In working on this topic, Einstein "feared he would be consigned to a madhouse." It is easy to understand why.

Einstein tried hard to find a simple, elegant solution for the cosmos. He strongly believed that the universe is static, so that nothing changes over time. There was no observational evidence to suggest that the universe was anything other than static. No matter how hard he tried, he could not find any such solutions to his equations. In 1917, he thought he had an answer. In looking over his equations, he found he could add a new term, one that became known as the *cosmological constant*. He included it in his paper titled "Cosmological Considerations in the General Theory of Relativity." Without the cosmological constant, there could be no static universe because gravity would cause it to collapse. The cosmological constant offset gravity by acting like a negative pressure. For decades, cosmologists wrongly thought that the cosmological constant had to be zero, and Einstein later called it the "greatest blunder of my life." Now, however, the best description of the universe we live in has a very nonzero cosmological constant. Perhaps Einstein's greatest blunder was thinking that the cosmological constant was his greatest blunder!

Shortly after Einstein's paper of 1917, the great Dutch astronomer Willem de Sitter found another solution to Einstein's equations. This event was important, for not only did de Sitter's universe evolve over time, but he communicated his conclusions to the Royal Society, the exclusive English meeting place for the world's greatest scientists. It was through de Sitter's letter to the Royal Society that the English-speaking world became familiar with general relativity. The secretary of the Royal Society, Arthur Eddington, a gifted astronomer, theoretician, and observer, became the biggest booster of general relativity in Britain. He knew that general relativity could be tested, and he thought the best test would be to see if starlight would be deflected near a massive object. The same idea had occurred earlier, independent of Eddington, to Einstein and Freundlich. Some experimentalists had already tried to detect a gravitational redshift, but the experiments were bogged down with technical difficulties and could easily be challenged. For the deflection of light, all they needed to do was look at light from a distant star as it passed close to the Sun. The trick, though, was to make the Sun dark so that you could see the light from that star. This meant that Eddington needed a total eclipse of the Sun. One was due in West Africa on May 29, 1919. Luckily, the Sun was in the constellation of the Hyades, the star cluster closest to the

Earth. Because it contains several bright stars, more than one star could be checked when the Sun was dark.

Einstein already knew the importance of this observation. An attempt to photograph a solar eclipse had already been made by Freundlich in August 1914 in the Crimea, but the outbreak of World War I had prevented it when the expedition members were arrested in Russia as enemy aliens. More than a year later, in December 1915, while corresponding with a colleague, Otto Naumann, about general relativity, Einstein wrote, "This consequence [the bending of starlight] is the most interesting and astonishing of all." Within a few weeks, he wrote to Schwarzschild that "the question of light deflection is now of the utmost importance."

The 1919 eclipse would prove to be more fruitful than the 1914 trip to the Crimea. The Royal Greenwich Observatory, through which the line of zero longitude passes, sent one team to Sobral, in northeastern Brazil. They took with them a 13-inch lens from their own observatory and, as a backup, a four-inch lens from the Royal Irish Academy. Eddington headed to the island of Principe off the West Coast of Africa, taking with him a lens from the Oxford Observatory. After taking photos of the sky during the eclipse, both teams had to stay on until the Hyades became visible in the night sky. That way, they could compare the nighttime photos with the eclipses photos and see any apparent change in the star positions.

More than a month later, on September 12, a meeting of the British Association for the Advancement of Science took place. The auditorium was packed. The star of the meeting was Arthur Eddington, who had come to announce his findings: the deflection of light lay somewhere between 0.83 and 1.7 arc seconds, the predictions made by Einstein in 1911 and 1915. Later, at a formal meeting of the Royal Society, held jointly with the Royal Astronomical Society, the Sobral team reported a deflection of 1.52 arc seconds as measured with the Greenwich lens but a 1.98 arc second deflection with the Irish Observatory lens. Given some problems they had had with the Greenwich lens, the result of 1.98 arc seconds seemed more reliable. The Principe group, on the other hand, reported a deflection of 1.6 arc seconds, still within an acceptable range. The results were a triumph both for Einstein and for British astronomy, but the physicist's response was fairly mild. When told that his theory was confirmed, Einstein replied, "I knew that the theory was correct. Did you doubt it?" When asked what he would have done if the observations had not agreed with theory, Einstein said he "would have to pity our dear God. The theory would be correct all the same." Max Planck had been more nervous than Einstein. When Einstein heard that Planck had stayed up all night, anxiously awaiting news of the results, he observed, tongue in cheek, that if

Planck "had *really* understood the general theory of relativity, he would have gone to bed the way I did."

It was not long before Eddington and Einstein were writing books on relativity. Eddington's *Space, Time, and Gravitation* appeared in 1920, and *The Mathematical Theory of Relativity* was published three years later. Einstein wrote *The Meaning of Relativity*, published in English translation in 1922. All three books would become international best-sellers, and the 1922 book is still in print. Close to his death, Einstein reworked the equations for *The Meaning of Relativity* in its fifth edition, prompting one of the local Princeton newspapers, no doubt recalling Planck's sleepless night, to write the headline, "New Theory Keeps Our Man Up Nights."

Years later, Princeton physicist Bob Dicke would question the results of the eclipse expedition. He realized that if the sun were slightly flattened and not as round as previously thought, it could account for the observed deflection of starlight. Master experimenter that he was, he set about constructing an experiment to see whether the sun is truly round or whether oblateness might play a role in the deflection. He found no oblateness, thereby helping to establish Einstein's theory more firmly.

Today we have another well-documented aspect of light bending. Astronomers now know that out in the universe, there are black holes, neutron stars, and pulsars whose gravity field is far more intense than that of our puny Sun. Instead of slightly changing the path of light rays, these objects bend light by a considerable amount and act as gravitational lenses. If a gravitational lens lies between us and a distant galaxy, the lens might bend light so much that we can see multiple images of the galaxy. Astronomers have found many examples of this phenomenon, which confirms Einstein's theory and assists them in the hunt for black holes.

* * *

After 1916, astronomers and mathematicians joined in the race to find new solutions to Einstein's equations. English mathematician Edward Arthur Milne found some simple solutions. So, too, did a Russian physicist, Alexander Friedmann. He found that a universe that expanded in time could exist. Einstein did not like this idea and tried to convince Friedmann that his mathematics must be wrong, but Friedmann stuck to his guns. Soon, Georges Lemaître, a Roman Catholic priest who was also an astronomer and one of Belgium's top bridge players, replicated Friedmann's idea. He realized that if the universe is expanding now, it must have been smaller in the past. There may have been a moment of "creation," when the universe had zero radius, and he called these early phases of the universe the "primeval atom." The work of Friedmann and Lemaître, made more mathematically formal by mathematicians Howard Robertson and Arthur Walker, was summed up in their *FLRW metric*, which is the most general form for the metric of a universe that is homogeneous and isotropic, one that has the same properties in all directions. It uses the cosmological principle—that the Earth, our solar system, and our galaxy do not occupy a special place in the universe. This has its roots in the sixteenth century, when a Polish monk, Nicholas Copernicus, suggested in his book *On the Revolutions of the Heavenly Spheres* that the Earth was not the center of the solar system. The expansion of Copernicus's idea—that we are in no special place at no special time—has become a powerful principle in determining which models of the universe are plausible.

Soon, other theoretical universes were proposed. For example, Kurt Gödel, a professor of mathematical logic at the Institute for Advanced Study in Princeton, came up with a doozy. He showed that Einstein's equations could be solved if the universe were homogeneous and static—an idea Einstein himself liked—but rotating. The *Gödel metric* exhibits completely off-the-wall behavior: it shows that time travel is possible. Physicists don't like such an idea because it sounds like science fiction and violates a cherished belief in causality—that an event happens only after something has caused it to happen. Cambridge University mathematician and cosmologist Stephen Hawking has proposed the "chronology protection conjecture" to eliminate all time-travel universes.

Other static solutions are possible as well. In 1948, a group of Cambridge-based astronomers proposed the *steady-state model*, which is static and homogeneous but continually adds matter into the universe. This approach helps avoid the nasty problem of having the universe packed into a dense region of spacetime right at the beginning of the universe. Fred Hoyle, one of the creators of this universe, did not particularly like the models of Friedmann, Lemaître, and others, mockingly calling their fireworks universe the "Big Bang" model. The name stuck.

So, is the universe static or expanding? In 1917, Vesto Slipher at the Lowell Observatory used the Doppler effect to show that many spiral galaxies were receding from us. Eddington had shown that Willem de Sitter's model for the universe, if it contained matter, meant that matter would expand. In the late 1920s and early 1930s, Edwin Hubble used the Mount Wilson Observatory in California to look at galaxies. He could detect that the distant ones appeared redder than those nearby, suggesting that the universe was expanding. If one galaxy was twice as far away as another, Hubble showed, it would recede twice as fast. This became known as *Hubble's law*. Years later, the Big Bang proponents claimed this observation supported their ideas, but the steady-state theorists showed that Hubble's data led to a problem. Hubble's numbers pointed to a universe that was a mere 1.8 billion years old, yet radioactive carbon dating of rocks showed that the Earth was at least 2 billion years old. Observational methods would improve over the years, and different techniques would be found, but by the early 1960s, the Big Bang model was the only viable candidate to explain the origin and current state of the universe.

One more stunning confirmation of general relativity needs to be mentioned. Lemaître had proposed that in its earliest phase, the universe behaved like a "primeval atom." But what would happen if there was radiation in the early universe? Bob Dicke, who had tested the principle of equivalence, felt certain that some relic of that radiation must be lurking undetected in the universe. He, together with his graduate students Jim Peebles and Dave Wilkinson, who jokingly referred to themselves as the "Dicke birds," set out to find it. One day, so the story goes, the phone rang in Dicke's lab at Princeton. Two scientists working for Bell Labs, Arno Penzias and Robert Wilson, had a problem with the radio telescope they had built: there was an infuriating background hiss when they tried to detect signals that had a wavelength of a few centimeters. Thinking the noise might be due to pigeon droppings, they cleaned out the equipment, but the hiss did not go away. Could Dicke help them? Dicke, ever the gentleman, covered the mouthpiece of the phone and said to Wilkinson and Peebles, "Boys, we've been scooped." Dicke went on to explain to Penzias and Wilson that the hiss probably meant they had discovered the remnant or radiation left over from the Big Bang. Penzias and Wilson wrote an article on their findings for the prestigious *Astrophysical Journal*. The very next article in the journal, "Cosmic Blackbody Radiation," was by Dicke and Peebles. These two Princeton physicists used Einstein's theory of general relativity to show that radiation from the Big Bang would have calmed down enough to have a temperature of about 3 kelvin. From Planck's law, radiation at 3 kelvin should have a wavelength of a few centimeters, which is why Penzias and Wilson picked up the signal—they were tuned in to the cosmic microwave background's frequency. Penzias and Wilson received the Nobel Prize in physics in 1978; Dicke, to the outrage of many physicists, received nothing.

Besides these, there was yet another noteworthy prediction of general relativity. Think back to the droplet hanging from the faucet or the basketball on the trampoline. If you tap the faucet or push the basketball down, the delicate balance between gravitation and the surface tension will be upset. Push the basketball down and let go: the trampoline's surface will gently move up and down in a wavelike motion. The droplet, if you tap the faucet, will have tiny ripples set up on its surface. The equations of relativity are similar. The curvature of spacetime is determined by the stress tensor of matter. If something drastic happens to the matter so that the stress tensor changes, then the curvature of spacetime will change. In a 1918 paper, Einstein predicted that gravitational waves could occur: if something dramatic happens to a star, it can cause ripples, or waves, in spacetime. For a normal star made of gas, the temperature of the gas produces a pressure that keeps the star from imploding under the force of gravity. As the star gets old, though, it runs out of the fuel it burns to keep warm. When this happens, there is nothing to offset the force of gravity, and the star implodes—that is, until it gets to the high-pressure core of the star. At that point, the outer layers that were headed inward bounce back out into space with a fantastic amount of energy, and we get a supernova explosion. Astronomers can often observe these explosions because of the enormous number of X rays they generate. Supernovas are also good candidates for generating gravitational waves. Gravitational-wave detectors are being built to hunt for them, which would be one of the great confirmations of general relativity.

In 1974, Americans Joe Taylor and Russell Hulse found a curious pair of stars. One of them was a pulsar, denoted PSR1913+16, sending out beams of radiation as it orbited an unseen companion, probably also a pulsar. The astronomers thought they might be seeing a binary pulsar system. In 1978, they showed that, according to general relativity theory, the system should emit gravitational radiation, which should alter the orbits of the pulsars. Taylor and Hulse observed that this is what happened. Over the next two decades, they watched the pulsars and, over the years, the pulsars' orbits changed. This was not just precession, as with the planet Mercury. The two astronomers showed that the change in the orbits meant the pulsars were losing energy. The energy was being emitted as gravitational radiation. Eventually, they were able to confirm that the stars did emit gravitational radiation in just the right amount that Einstein's theory predicted. They won the Nobel Prize in 1993.

* * *

General relativity was completed by 1916. It is now part of the training of every physicist. The Big Bang model of the universe is entirely in harmony with general relativity, and a new generation of astronomical observatories is providing high-tech confirmation of Einstein's ideas. The microwave background radiation can be explained by relativity. High-resolution maps of the microwave background, produced first by the Cosmic Microwave Background Explorer and now through the Wilkinson Microwave Background probe, named for Dicke's former student Dave Wilkinson, have helped find tiny fluctuations of the radiation. General relativity has thus become an applied science.

Chapter 8

THE LATER BERLIN YEARS: POSTWAR TURMOIL AND THE RISE OF HITLER

> I have now been promoted to being an evil monster in Germany, and all of my money has been taken away. But I console myself with the thought that it soon would have been spent, anyway.
>
> —To Einstein's friend, German physicist Max Born, May 30, 1933, after the money in his Berlin bank account had been confiscated

Prompted by "mysterious wireless signals" received from an unknown source in both London and New York, a London correspondent contacted Einstein for an explanation. In an interview on "interplanetary communication" in the *Daily Mail* of London at the end of January 1920, Einstein said there is every reason to believe that Mars and other planets are inhabited, but Martians would be more likely to communicate via light rays than through the wireless. He hypothesized, however, that the signals were due either to atmospheric disturbances or to secret experimentation of other systems of wireless telegraphy.

Those are the sort of questions Einstein now took time to answer, and the fact that he answered them at all made him even more popular around the globe. After the results of the 1919 eclipse expedition were broadcast worldwide to a war-weary public, his fame rose, and he became the first international scientific celebrity. People from all walks of life wrote to him, whether they were children or pensioners, rich or poor, scientists, or simply people wanting to have contact with someone who was now considered a true genius. He felt like King Midas: "Just as with the man in the myth who turned whatever he touched into gold, with me everything is turned into newspaper clamor," he wrote to physicist Max Born in September. He also felt his achievements were overrated and he did not deserve the "boundless admiration" he was receiving. Because the theory was so difficult to understand and explain in nonscientific terms, it became all the more mysterious and added to the myth that only Einstein could see a dimension to the universe that was beyond almost anyone else's comprehension. He enjoyed his sudden fame and was jocular with reporters; eventually, however, he began to retreat from the commotion that would not subside until the end of his life—and not even then.

Earlier in the year, Einstein's mother, Pauline, had died of stomach cancer at the age of 62. Einstein had been worried about her for several months while she was living in Switzerland. He spent much time arranging to bring her to Berlin and giving her a room of her own in his apartment so he and Elsa could care for her properly and lovingly during her last days of life. After her death, he wrote to his friend Heinrich Zangger in Switzerland, "My mother died a week ago today in terrible agony. We are all completely exhausted. One feels in one's bones the significance of blood ties." To another friend, Hedwig Born, he wrote, "I know what it's like to see one's mother go through the agony of death without being able to help. ... We all have to bear such heavy burdens, for they are unalterably linked to life."

In the early 1920s, Einstein became aware of increasing anti-Semitism as nationalism continued to soar after the defeat of Germany during the war. Signs of anti-Jewish bias were emerging in Berlin against Jews in general and against Einstein, the new celebrity, in particular. Part of the prejudice appeared in the form of outrageous criticisms against relativity theory—which anti-Semites, particularly German physicists Philipp Lenard and Johannes Stark, both of whom had won the Nobel Prize, referred to as "Jewish science." At first, Einstein tried to ignore and dismiss their remarks, preferring to involve himself with the concerns of reconstruction after the war. Later, with the rise of Nazism, he faced the seriousness of the state of affairs and became much more vocal.

Right now, however, his focus was on education. Einstein strongly believed in equal opportunity in education for anyone who wanted it. He displayed this belief in his liberal admission policy to his classes both while a visiting lecturer in Zurich in 1919 and again while teaching at the University of Berlin in 1920. In February 1920, the student council in Berlin protested his open-admission policy on behalf of the paying students and heckled him during a lecture. Some claimed the heckling was due to anti-Semitism because many unregistered "students" in the audience appeared to be Jews from eastern Europe who had immigrated to Berlin and were anxious to hear Einstein speak. But the student council, administration, and Einstein himself denied this was true. Because of this "uproar," as the headlines in Berlin's newspapers that month called it, which continued for several days, the administration changed its policy. However, soon after this event, Einstein reversed his stand slightly by saying the unregistered public should be allowed into the lecture hall if seats were still available after the paying students were seated. But just in case all seats were taken, he also arranged to give a series of free evening lectures outside the elite university.

This year, Einstein made lecture trips to Holland, where he was given an appointment as visiting professor; to Norway; and to Denmark, where he met with physicist Niels Bohr, whom he had already met during Bohr's visit to Berlin in February. He also published statements on science for the general public. In one, solicited by the Association for Popular Technical Education in July 1920, he stressed the importance of showing to students and the public the practical applications of science and technology to everyday life and stated his belief that a technical education is equal in value to a humanistic one.

In the meantime, Adolf Hitler, an unknown organizer for a German nationalist party, the German Workers' Party, began to establish himself in politics. In February 1920, Hitler organized a large event for a crowd of nearly 2,000 in the Munich Hofbräuhaus and presented a 25-point program that would define his party. The party was soon renamed the National-Socialist German Workers'—or Nazi, an abbreviation of the German Nationalsozialist—Party, and it ushered in an era of fascist extremism in Germany lasting 25 years.

By August, Einstein felt it was time to repudiate the attacks made on his theory since 1918 by what he called "the Anti-Relativity Company." After he delivered two lectures in the Berlin Philharmonic Hall, his accusers charged him with plagiarism, with being hungry for publicity, manipulating the press, conducting un-German science, and so on. Outside the hall, the organizers distributed anti-Semitic leaflets and sold swastika lapel pins. He responded to them angrily in a newspaper statement as well as in a lecture a month later after having desisted from making public statements on earlier provocations. These encounters polarized the German physics community, and Einstein's supporters were afraid it might drive him out of the country.

Around this time, Einstein also wrote two documents on the "Jewish question" to reflect his growing awareness of the growth of anti-Semitism after the war. Even though he did not publish the articles, he spoke out on the subject. His statements were not always popular with the Jewish citizens of Germany and Switzerland. In one instance, for example, Einstein had been invited to take part in a meeting dedicated to fighting anti-Semitism in academic circles. He declined to take part because he thought such an endeavor would not prove fruitful. He maintained that Jews must first fight against the anti-Semitism existing among themselves (that is, western European Jews against eastern European Jews), and only when they respected themselves would they be able to earn the respect of others.

In the spring of 1921, Einstein decided to make his first trip to America. He had two objectives. The first was to take part in a fund-raising tour on behalf of a Hebrew university the Zionists were planning to establish in Jerusalem. He strongly supported the founding of the university, favoring an emphasis on the sciences and health professions. He felt a Hebrew university would give Jews who had been barred from universities in other nations a place to study, teach, and conduct research. His second objective was to visit Princeton University in New Jersey, whose officials had invited him to deliver a series of four lectures on relativity theory. Because of his poor health, he insisted that Elsa accompany him. Leading the trip on behalf of the Zionists and Hebrew University was future Israeli president Chaim Weizmann, the new president of the World Zionist Organization and a British citizen. The two men apparently discussed not only Zionism and politics but also science. "During our crossing, Einstein explained his theory to me every day, and by the time we arrived I was fully convinced he understood it," joked Weizmann.

On his arrival in New York on the Dutch ocean liner *Rotterdam*, Einstein immediately became popular. Through an interpreter, he bantered with the reporters who had rushed on board the ship to be the first to see the genius. Afterward, city officials took him on a motorcade down the streets of lower Manhattan to City Hall, where Mayor James Hylan welcomed him. To packed auditoriums, during his stay he proceeded to give lectures at Columbia University and the City University of New York. Then, before heading to Princeton, he delivered a speech to the National Academy of Sciences in Washington, D.C. Fifty-eight years later, a huge and popular statue of him was placed in front of the Academy building to honor the centennial of his birth. All his lectures, delivered in German, were translated for the audience. While in the nation's capital, Einstein, with members of the National Academy of Science, briefly went to the

White House to meet President Warren Harding after Harding's initial rejection of the idea. Einstein's reception in Washington, which apparently had not yet warmed up to Germany three years after the end of war, was rather cool. French Nobelist Marie Curie, on the other hand, was treated more ceremoniously two weeks later.

In Princeton, Einstein kept busy as well. Besides delivering four lectures in McCosh Hall's huge lecture room 50, he also accepted an honorary doctorate. "We salute the new Columbus of science voyaging through the strange seas of thought," declared Dean Andrew West after he read Einstein's citation, before President John Hibben formally conferred the degree on the physicist. Part of Einstein's motivation for the Princeton lectures was financial—he was strapped for money because of his many family obligations, and he was looking for financial resources from abroad during the difficult postwar economic situation in Germany. Princeton University had originally invited him to give a series of three lectures a week over a two-month period, for which he asked $15,000 in payment, a huge sum for that time. As a compromise, the series was cut down to four lectures delivered within a week called "The Meaning of Relativity," for which he received a much smaller fee. A book based on these lectures was published the following year both in the United States and in England and is still in print. After fulfilling his obligations on the East Coast, Einstein continued on to the Midwest and Chicago to complete his fund-raising effort with Weizmann. Even though this part of Einstein's trip was not as successful as the Zionists had hoped, the physicist seemed to solidify his Jewish identity during this time as he mingled with the Jewish communities in America. "Zionism really represents a new Jewish ideal," he wrote to his friend Ehrenfest after his return to Germany, "one that can give the Jewish people once more joy in their existence. I am very glad I agreed to Weizmann's invitation."

Before returning home, the Einsteins had stopped in England, where Albert gave a lecture at King's College at the University of London on the theory of relativity and another at Manchester University, where he signed his name on a blackboard that is still preserved. He was supposed to receive the Gold Medal of the Royal Astronomical Society at this time, but it was rescinded because Einstein was a German; like the Americans, the British had not yet forgiven Germany for its wartime transgressions. In an effort to endear himself to the citizens of London and to foster German-British postwar reconciliation, Einstein paid his respects at the tomb of Sir Isaac Newton in Westminster Abbey, leaving a bouquet of flowers there.

After his trip to the United States this year, Einstein made some remarks about Americans that were not popular back in the States, especially one in which he characterized American men as the lapdogs of their wives and wives as extravagant spenders of their husbands' money. He later tried to explain, if not rebut, his statements, commending Americans on their warmth and friendliness and expressing admiration for the close relationships between students and teachers in schools and universities. He noted that American patriotism is not so much nationalistic as an inner pride of being citizens of a great country.

Soon after his return to Berlin, Einstein took time out to take a vacation with his sons, Hans Albert, who was now 17, and Eduard, who was 11. They went sailing on the Baltic coast up north, staying overnight in a room in the local village bakery. Einstein preferred this kind of simplicity much more to the more elaborate accommodations he had with Elsa on his American and English trips. He was getting along well with the boys now and admitted Mileva was doing a good job in raising them. He also involved them in designing a gyrocompass for a large German company.

Though Einstein did not receive the official news until the following year, toward the end of the year he heard that he would receive the Nobel Prize in physics for 1921 for his work of 1905 on the photoelectric effect, which led to the communications revolution. Because some physicists were still disputing relativity theory, the Nobel award did not mention the work that had made him most famous. Receiving the prize also brought the question of his citizenship to the fore. Einstein had considered himself a Swiss citizen until this time, but he was advised by the German government that, along with his appointment to the Kaiser Wilhelm Institute in 1914, he had received automatic German citizenship. By this time, Einstein had no objection to this status as long as he could continue to keep his Swiss citizenship as well, to which the German authorities agreed.

Meanwhile, the German mark plummeted, and a recession set in. Most of the economic difficulties of the postwar period have generally been blamed on the harsh conditions imposed on the Germans by the Treaty of Versailles, the peace agreement between Germany and the Allies. The reparations demanded by the European nations were draconian, and it fell to ordinary German citizens to pay the debts. The government tried borrowing and printing more money, resulting in an extremely high inflationary period and contributing to political unrest on all sides of the political spectrum, from Communist to fascist. In the end, the fascists, the most nationalistic of the groups, were favored because they denounced the Versailles Treaty most strongly and opposed the democratic goals of the new German republic. With the rise of fascism came the rise of Hitler and his Nazi Party, which took advantage of and manipulated the Germans by promising reforms that would end both their humiliation and their desperate economic situation. Soon, with the increasing rise of nationalism, the stage would be set for another world war.

In their despair during the interwar period, Germans of the new Weimar Republic, as the government was referred to between 1919 and 1933, became engaged in a strange renaissance in experimental new forms of art, music, and literature. Jazz, classical music, cabaret, theater, surrealism, and the Dadaist art movement and literature all found lasting expression at this time, especially in Berlin, which drew artists and the curious from all over the world. By 1923, a system of financing with international loans was arranged that worked well for Germany, reducing inflation until the financial systems broke down worldwide in the early 1930s.

Einstein was too busy with physics and his more serious concerns to engage much in the hip new scene. In an article expressing his views on education in 1921, "Einstein on Education," for instance, he concluded that present-day mathematical and scientific education is too concerned with abstractions and should be more hands-on, that testing students with exams should not be necessary (something he felt strongly about all of his life), and that the school day should last a maximum of six hours. In another statement, he emphasized that art and science have common elements in that both give expression to creativity: in science through logic and in art through form, both of which help one escape from merely personal concerns.

In 1922, as Wilhelm Cuno became chancellor of Germany, Einstein completed his first paper on a "unified field theory." In this comprehensive theory, Einstein attempted to express gravitational theory and electromagnetic theory within a single unified framework to show that the two forces are based on one grand underlying principle. He had trouble accepting two distinct forces in nature—and this was even before the discovery of two additional forces: the strong and weak nuclear interactions. Today, physicists are searching for an even loftier theory: a Theory of Everything (TOE), which they hope will unify all the forces and all matter. Currently, string theory—developed originally at Princeton University by John Schwartz, Ed Witten, and David Gross and at Queen Mary College of the University of London by Michael Green during the early 1980s—appears to be the tool of choice.

Einstein now became interested in international political activities. With Marie Curie, H. A. Lorentz, and others, he joined the newly formed League of Nations' Committee on Intellectual Cooperation, whose goal was to mobilize intellectuals from around the world to work for peace. He soon became disaffected with the League, however, believing it did not have the "sincere desire" to achieve its aims, and quit, though he did not renounce its principles. A year later, he changed his mind and rejoined, attending meetings regularly until 1930, when he withdrew permanently. His withdrawals were due mostly to his general unsuitability for committee work since he preferred to work mostly on his own to achieve his goals. All the while, Germany was continuing to experience economic hardship and depression, and Communism was rearing its head in the east, gaining special popularity among the poor. Jews seemed a good scapegoat to many for both of these situations. Anti-Semitism became ever more widespread.

Most Germans, however, continued to celebrate Einstein. They wanted the prestige of his name to add gloss to their institutions even as he criticized them. He even traveled in some of these circles but always remained friendly, informal, and "democratic" toward all, believing one should respect honest people even when they have views different from one's own. However, after two nationalist fanatics assassinated the newly elected German foreign minister, Walter Rathenau, a Jew and a friend of Einstein's whom the emerging Nazis accused of being part of a "Jewish-Communist conspiracy," in June 1922, Einstein took heed and began to remove himself from any controversial political arenas. He was warned, for the sake of his safety, not to make any public appearances. He wrote to his friend Maurice Solovine on July 16, "There has been much excitement here ever since the abominable murder of Rathenau. I myself am being constantly warned to be cautious, have canceled my lectures, and am officially 'absent,' although actually I have not left." But he did begin to think about resigning from his position at the Kaiser Wilhelm Institute and moving away.

First, however, came the opportunity to take another extended trip abroad as well as a temporary chance to escape from possible harm at home. Einstein was invited to Japan, and he planned to make stops in Palestine, Spain, and France on the way back home. He and Elsa were the guests of a Japanese publisher, Kaizosha, which offered him a generous fee for a series of lectures throughout Japan. He later wrote his "Impressions of Japan" for the magazine *Kaizo*. The Einsteins left Germany at the beginning of October and headed to Marseilles in France to board the Japanese steamer *Kitanu Maru* for what would be a six-week excursion to Tokyo. Along the way, they stopped in Singapore, Hong Kong, Ceylon, and Shanghai. Einstein took along a notebook and kept a travel diary of his trip, writing his impressions of the people and places that were so new to him.

Still on board ship, Einstein once again developed severe abdominal pains, and a doctor gave him the required medical attention. Later, while in Hong Kong or en route to Shanghai, he received the news that no doubt made him feel much better: his Nobel Prize award was now official and had been announced. Now, with the added prestige of being a Nobel laureate, he proceeded to give lectures in Tokyo, Sendai, Kyoto, and Fukuoka. He was received and treated like an international celebrity everywhere he went and was touched by the warmth and adulation of his hosts, admiring "their refined customs and lively interest in everything." He wrote to his friend Besso in May 1924, "I have for the first time seen a happy and healthy society whose members are fully absorbed in it." In his last public event, in Fukuoko, to the great delight of the audience, he played his violin at a YMCA Christmas party. A few days later, shortly before New Year's Day 1923, he and Elsa left the Far East and departed for the long voyage to the Middle East.

A month and many memories later, they were on the shores of Palestine, which had been under British rule as a mandate since World War I. The things Einstein saw and experienced here would leave a deep and lasting impression on him. Here, in Tel Aviv and Jerusalem, Jewish immigrants from all over the world were working as artisans, skilled laborers, and farmers displaying their practical skills. Seeing them made him less skeptical than he had been earlier about the ability of Jews to build and develop a new homeland. Einstein was not so captivated, however, by the followers of the Orthodox Jewish tradition who spent much of their time in prayer, carrying on ancient traditions that were alien and distasteful to him.

Within a few days, Einstein was taking part in elaborate and historic ceremonies marking the foundation of the Hebrew University of Jerusalem. After giving a speech befitting the occasion—he read the first few sentences in Hebrew, a language he did not know, then continued in French—he laid the cornerstone of the campus's first building. In an article he later wrote about his view of the future Jewish homeland, Einstein had not yet shown any concern about an "Arab question." He declared that Jews and Arabs appeared to live in harmony and that the major problems were those of sanitation, malaria, and debt. For now, he saw the future homeland as a "moral center" for the Jewish people, though he was doubtful that the land could absorb all the Jews who wished to live there. Although he continued to be wholeheartedly involved with the establishment of Israel throughout his life, Einstein did not have the opportunity to return there in later years to see the fruits of his efforts and those of others.

By the end of the decade, after major anti-Jewish disturbances had occurred in Palestine, Einstein became more concerned about the Arabs and their rights. His attitude toward Zionism began to change, too, and the direction in which the Hebrew University officials were heading was not what he had envisioned when he had endorsed the project. Most of the money for the university had come from American Jewry, who felt they were now entitled to determine the goals of the new institution. While Einstein envisioned a first-rate teaching and research university dedicated to the highest scientific standards, one that would take into account the needs of Jews coming to Palestine, the Americans were content with a teaching college that would favor and employ wealthy American contributors rather than the most qualified scholars. He therefore soon found himself at odds with the founders. He felt the leaders were weak and chose "morally inferior men" to accomplish their goals. After some bitter exchanges with university officials and unwilling to compromise his standards and vision, Einstein quietly resigned his positions on the board of governors and the academic council.

With regard to the Arabs, Einstein felt Jews should find ways to reconcile and cooperate with them. Increasingly feeling the effects of Jewish immigrants encroaching onto their land, the Palestinians were now compelled to fight back. "Should we be unable to find a way to honest cooperation and honest pacts with the Arabs, then we shall have learned nothing from our 2,000 years of suffering and will deserve our fate," he wrote to Chaim Weizmann toward the end of 1929 in an appeal that Zionists cooperate peacefully with the Arabs. He hoped "the two great Semitic peoples" would have a great common future. His friend, number theorist Ernst Straus, summed up Einstein's views on Zionism well: "He was a Zionist on general humanitarian grounds rather than on nationalistic grounds. He felt Zionism was the only way in which the Jewish problem in Europe could be settled. ... He was never in favor of aggressive nationalism, but he felt that a Jewish homeland in Palestine was essential to save the remaining Jews in Europe." Einstein did not favor an independent Jewish state, but rather a land for both Jews and Palestinians governed by some international organization.

But for now, invigorated by his feelings of unity with his Jewish "tribe," as he called them, Einstein, with Elsa, left Palestine and traveled on to France and Spain before returning to Berlin in the middle of March. Einstein's visit to Spain had stimulated a national debate on the nature and social value of science, which was unusual for a country that had just recently begun the process of modernization. As elsewhere, Einstein's universal appeal cut across social classes and professions and opened people's eyes to the value and applications of science.

In July, Einstein traveled to Göteborg, Sweden, to deliver his Nobel lecture during a session of the Scandinavian Scientists' Convention. Because he delivered the lecture several months after he received the award, he did not discuss the prize topic—the discovery of the photoelectric effect. By now, scientists were more eager to hear Einstein lecture on the hottest topic of the day, so he surveyed relativity theory instead. Among those in the audience of 2,000, the large majority of them not physicists, was the king of Sweden, Gustav V. To satisfy those interested in a more technical lecture, Einstein later spoke about his unified field theory to a smaller audience of scientists.

After the prize money for the Nobel award was duly deposited directly into a Swiss bank account, Mileva and Hans Albert complained that they did not have free access to the full amount and could withdraw only the interest. Einstein, surprised by their attitude, felt he had already been more than generous with them. Later, on the suggestion of their advisers, the family decided to buy three rental properties that would provide Mileva with a lifelong income. That summer, Einstein reconciled with the family and spent time with his sons vacationing in southern Germany.

Einstein now returned to promoting his pacifist goals. He pleaded that because technical inventions that arise from science have international consequences, including military applications, men must create organizations dedicated to preventing wars in which such products might be used violently. He joined—or at least lent his name to—several left-wing organizations dedicated to liberal social and political causes. These affiliations made him a natural target for right-wing extremists who made threats on his life. The threats were probably connected to Hitler's Beer Hall Putsch (attempted coup d'état) in Munich in November 1923, at which Hitler had hoped to make plans to overthrow the new Weimar Republic. The putsch failed, and Hitler was arrested and put on trial. He was found guilty and sentenced to five years in prison, though guilt for attempting to overthrow the government would normally have meant the death penalty. Clearly, the judges who sentenced him sympathized with his goals. Claiming to be depressed and threatening to commit suicide, Hitler was released by sympathetic Nazi officials in the Bavarian government for "good behavior" after an incarceration of only eight months. While in prison, he wrote the first part of *Mein Kampf* (My Struggle), in which he expressed his hatred of Jews and Slavs and berated democracy, and was a favorite of the jailers. Einstein, meanwhile, fled to the Netherlands for a short time for safety until the worst of the unrest subsided.

By the middle of May 1924, Einstein was able to write to a friend that political conditions in Germany had become less explosive and that his life had become tranquil and undisturbed. The peaceful environment was a good time for the occurrence of a happy family event. Einstein took on a new role—as stepfather-in-law. Elsa's elder daughter, Ilse, was married to Rudolf Kayser, an editor and publishing adviser. In 1930, Kayser would publish the first biography of Einstein under the pen name of Anton Reiser. Hoping to make the biography part of Einstein's fiftieth birthday celebrations in 1929, Kayser missed the deadline by a year. Einstein disliked biographies about himself, feeling that such publications would make him look vain and publicity hungry. He never read anything that was written about him: "In this way one doesn't get spoiled by praise or depressed by blame," he confided to a friend much later. The only biographer he tolerated was Carl Seelig, though he did not read his biography, either; to get around his vow not to read it, his secretary, Helen Dukas, read parts of it to him instead.

After Ilse, whom he had five years earlier proclaimed to love, became unavailable for his romantic attentions, Einstein became enamored with his new secretary, Betty Neumann. He ended the relationship by the end of the year, knowing it could never lead anywhere. Later on, even while he was still married, a succession of other women would enter Einstein's life. Not only was he world famous—a scientific rock star—but he was also handsome in midlife and charming and youthful. He was heavily muscled and unusually strong, and those who met him often remarked on the "twinkle" in his eyes. Einstein's humor and laughter were also legendary. "He responded with one of the most extraordinary kinds of laughter. It was rather like the barking of a seal. It was a happy laughter. From that time on, I would save a good story for our next meeting, for the sheer pleasure of hearing Einstein's laugh," recalled one of his assistants, Abraham Pais, later.

Einstein had affairs with some of the women who availed themselves to him and ignored others. His favorites were those with whom he could share intellectual and musical events or who enjoyed sailing in his modest sailboat. "Sailing is the sport that demands the least energy," he claimed. "But I am not so talented in this art, and I am satisfied if I can manage to get myself off the sandbanks on which I become lodged." Often he would even escort the ladies to glamorous events with Elsa's knowledge and deference. His marital relationship with his good-natured wife was grounded on companionship rather than passion, and there is no indication that she had any violent reactions or was jealous about her husband's dalliances. More than likely, she resigned herself to these short-term affairs and took pleasure in being the wife of a famous man, confident that he would never leave her.

After having declined an invitation to visit and lecture in South America in 1922, Einstein in 1925 accepted a renewed offer to go there. In March, he embarked on a three-month trip to Argentina, Uruguay, and Brazil, again taking a travel diary with him to record his impressions. He was welcomed and feted on the continent by the countries' scientific organizations, politicians, and their German and Jewish communities and gave lectures as required by his hosts. For Einstein, his weeks there were relaxing and "great fun."

As noted earlier, beginning in the mid-1920s, as Einstein was facing middle age, his concerns became more political and social and less scientific. Though he still worked on and published papers on quantum mechanics and reported on his unified field theory, which was hitting something of a dead end, he was no longer making major contributions in physics. This decline in productivity with age is not so unusual in the physics and mathematics professions; most physicists and mathematicians make their most original and creative contributions earlier in their careers. One of Einstein's scientific friends would remark later that, after around 1925, "Einstein might as well have gone fishing." Still, because of his accomplishments and charismatic personality, his reputation as the greatest living scientist did not diminish. Fellow physicists would send him their papers for evaluation. For example, many years later, cosmologist George Gamow, after receiving a letter from Einstein stating that one of his ideas is probably correct, scribbled on the bottom of the letter, "Of course, the old man agrees with almost anything nowadays." Professional organizations continued to honor him for his achievements. Within only two years, between 1926 and 1928, the Royal Astronomical Society of England awarded him its Gold Medal; Britain's prestigious Royal Society, of which Newton had once been president, awarded him the Copley Medal, its highest award given annually, alternating between the physical and the biological sciences; and the Academy of Sciences of the Soviet Union made him an honorary member.

Einstein channeled part of his energy into pacifist activities, which he had mostly given up during World War I because he thought any further effort was useless at that time. After the war, he rekindled his efforts to prevent future wars. In 1922, for example, he had contributed to a handbook on pacifism in which he said that scientists favor pacifist goals because they need internationally organized cooperation to carry on their work. Furthermore, "No person has the right to call himself a Christian or Jew so long as he is prepared to engage in systematic murder at the command of an authority, or allow himself to be used in any way in the service of war or the preparation for it." In 1925, Einstein, affirming his pacifist convictions, signed a manifesto against compulsory military service. Among the other signers were Mahatma Gandhi, the philosopher, pacifist, and leader of India's independence movement who was famous for his advocacy of nonviolence as a means of revolution; Indian mystic, poet, musician, and artist, Nobel laureate Rabindranath Tagore; and the English writer H. G. Wells, perhaps best known now for his short stories and science fiction novels, among them *The Time Machine* and *The War of the Worlds*. Einstein felt that war can be abolished only if a worldwide resistance against military service is organized internationally. "There are two ways of resisting war," he wrote in 1931, "the legal way and the revolutionary way. The legal way involves the offer of alternative service not as a privilege for a few but as a right for all. The revolutionary view involves an uncompromising resistance, with a view to breaking the power of militarism in time of peace or the resources of the state in time of war."

In 1925, Germans elected Paul von Hindenburg as president, and Hans Luther became chancellor. Hindenburg had retired from the German army in 1918, but he continued to take an active interest in politics. He was reelected president in 1932 and did not oppose the rise of Hitler, whom he appointed chancellor in January 1933. Hindenburg was so popular with the German people that Hitler was unable to overthrow the constitutional government until the president died in 1934. In the intervening years, the Hitlerjugend fascist youth organization was founded, Hitler's *Mein Kampf* was published, and Germany was admitted to the League of Nations. Paul Joseph Goebbels became the Nazi Party leader in Berlin.

In the fall of 1927, the Solvay Congress was again held in Brussels. The world's most notable physicists again attended it, this time to discuss the newly formulated quantum theory. The stars of this conference were Einstein and Niels Bohr, and throughout the conference they sparred over quantum mechanics. The debate would continue until Einstein's death. The year before, in 1926, in a letter to physicist Max Born, Einstein had written, "Quantum mechanics is very worthy of regard, but an inner voice tells me that this is not yet the right track. The theory yields much, but it hardly brings us closer to the Old One's secrets. I, in any case, am convinced that He does not play dice." He wrote something similar much later, in 1942, to another friend: "It is hard to sneak a look at God's cards. But that he would choose to play dice with the world is something I cannot believe for a single moment." On hearing Einstein's pronouncements on God, Bohr is said to have told Einstein, "Stop telling God what to do!"

Hans Albert, at the age of 23, announced his intention this year to marry Frieda Knecht, a woman nine years his senior. Einstein was opposed to the marriage, and even Mileva felt that Frieda was not good enough. She wrote to a friend the following year, after the marriage had already taken place, saying Hans Albert looked "frighteningly terrible" and "his wife does not know how to look after him, she thinks only of herself." Frieda, who was in fact a warm and bright woman, bore three children—David, Klaus, and Bernhard—only one of whom, Bernhard, lived beyond the age of six. Hans Albert and Frieda adopted a girl, Evelyn, in 1941.

The two-hundredth anniversary of Isaac Newton's death in 1727 was celebrated worldwide this year, and Einstein was among those asked to write tributes to the man on whose shoulders he said he stood. He also took the opportunity to praise the British for their traditions and for providing an atmosphere over generations that allowed the human soul "to soar." He acknowledged that everything that happened in theoretical physics since Newton's time developed from Newton's ideas and that only in quantum theory was Newton's differential method inadequate.

This year, in February 1928, Einstein's old mentor, Dutch physicist H. A. Lorentz, died. Einstein rushed to Leyden to deliver a celebrated eulogy over Lorentz's grave in which he called Lorentz the "greatest and noblest man of our time." "He shaped his life like a precious work of art down to the smallest detail. His never-failing kindness and generosity and his sense of justice, coupled with a sure and intuitive understanding of people and human affairs, made him a leader in any sphere he entered." He later said that Lorentz had meant more to him than anyone else he had encountered in his lifetime.

In March 1928, while on a trip to Davos, Switzerland, Einstein collapsed with a serious heart condition. Confined to bed for four months, it took him a year to recover fully. He now needed someone capable and dependable to take care of his growing personal affairs and correspondence, for which Elsa had no time. Therefore, with Elsa's help and approval, he hired a young woman, Helen Dukas, as his secretary. Helen was largely self-educated, having left school at 15 to take care of her family, and she remained with Einstein as part of his household for the remainder of his life. The two were able to establish a harmonious working relationship that lasted 27 years without getting romantically involved. Fiercely loyal to him, she was largely responsible for protecting and projecting his image as a lovable, kind, and benevolent, absentminded genius and for keeping his sometimes rakish personal life private. After his death, she became the dedicated archivist of his collected papers at the Institute for Advanced Study in Princeton—a job that gave purpose to her life in old age—until the archive was moved to Jerusalem in late 1981. A bright, shy woman with a sharp wit, she may be best known for her remark to physicist Ernst Straus after she was reintroduced to him in Princeton: "Of course I know you well. I was present at your circumcision." She died within a month after the transfer of the archive. She was 85.

Among Einstein's many obligations that had necessitated a full-time secretary in 1928 was his new position on the board of directors of the German League for Human Rights, a pacifist organization. As in 1917, he was again too ill again to be active for several months; he was also unable to attend the meetings of the Prussian Academy of Sciences. His colleagues would stop by his apartment and pick up his papers and speeches, which they would deliver on his behalf. In them, he presented a new mathematics that he felt would allow him to formulate his unified theory of gravity and electricity, which he continued to pursue doggedly.

On March 14, 1929, Einstein celebrated his fiftieth birthday. He was showered with congratulatory cards, letters, and telegrams sent by young and old from all over the world. In honor of this occasion, his friends were anxious to obtain a substantial gift for him: a summer home to call his own, which they hoped would be donated by the city of Berlin. However, the plan was foiled by politics and property rights. Bemused by the goings-on but tired of witnessing one fiasco after another, Einstein decided to buy his own plot of land and build a summer home in Caputh, out in the countryside near Berlin, three minutes from a lake where he could sail. His friends, embarrassed by the botched house-buying plan, surprised him with the next best thing: a sailboat of his own to sail on the lake. The house was built quickly that summer, and the Einsteins occupied it in September, keeping the apartment in Berlin as well. Einstein exulted in the rural setting, and his health improved at a faster pace.

With renewed energy, he settled into carrying on an extensive series of interviews and correspondence on behalf of Zionist and pacifist organizations. Most important, however, he published a paper, "On the Unified Field Theory," in 1929 that was a sensation even before publication. Gossip in the mass media had claimed that Einstein had solved the riddle of the universe, and throughout Germany the publication of his conclusions was anxiously awaited. The first printing of the paper sold out immediately, and more printings were ordered. Einstein could not understand the commotion. Even though he had faith in his theory, he didn't think his findings were as sensational as the public was led to believe. It was a classic case of media hype, and soon enough the critics found fault with his theory. After a few more attempts at corrections, Einstein finally had to admit, three years later, that his conclusions were wrong.

At this time, Einstein also began a friendship and long correspondence with Queen Elisabeth of Belgium and with her husband, King Albert. (He died in 1934.) On a visit to his uncle, Caesar Koch, his mother's brother, Einstein was invited to visit the queen, a lover of music and intelligent conversation, in her palace at Laeken. At the queen's request, they played music together in a trio, with a lady-in-waiting as part of the group. Over tea, the queen charmed Einstein with her modesty and cordiality, and he visited her and her husband again a year later. Theirs became a warm and enduring friendship, one that became extremely important to Einstein only a few years later when he sought refuge from Nazism. Einstein corresponded with the queen until the end of his life, always beginning his letters to her with "Dear Queen."

In 1930, family events took center stage. Einstein became a grandfather for the first time when Hans Albert's wife, Frieda, gave birth to a son, Bernhard; and he became a stepfather-in-law again when Margot tied the knot with Dmitri Marianoff (the marriage later ended in divorce). Einstein's other stepson-in-law, Rudolf Kayser, succeeded in publishing his biography of Einstein that year; Marianoff would publish his own version 14 years later. During the year, Albert and Elsa prepared themselves for another trip to America, this time to California, where Einstein was invited to lecture at the California Institute of Technology in Pasadena.

In the meantime, Einstein continued to step up his political activities in the face of increasing German nationalism and unrest. The Nazis had gained a majority in the German elections, portending as-yet-unknown evils. Even though right-wingers had taken over the national parliament, Einstein was not yet overly concerned about anti-Semitism, feeling that the reasons for the victory were economic rather than anti-Semitic. Because of the worldwide depression, including the one in the United States, the German economy was on a downward spiral again, many people were unemployed, the times were hard, and Germans hoped that a new government would bring positive change.

In the Soviet Union and its satellites, however, anti-Semitism was growing conspicuously. Consequently, Palestine now, in the 1930s, began to receive huge waves of eastern European Jews—about 200,000—arriving from the Russian shtetls to resettle the land of their ancestors. The British were aware of Arab resistance to such a huge influx and recommended the cessation of Jewish immigration. At this time, about 9.5 million Jews lived in Europe, mostly in the cities; those in Germany, numbering around half a million in a German population of about 65 million, were the most assimilated in Europe. For the most part, they had no desire to join the early exodus.

That summer, Einstein agreed to meet with Indian artist, musician, and poet Rabindranath Tagore at the Einstein's new home in Caputh. Tagore, with his exotic-looking entourage in tow, arrived in his plain long flowing robes, his considerable white hair and beard combed, looking every inch the mystic he was purported to be. In a set of two conversations, one in Caputh and the other at the nearby home of a friend, Bruno Mendel, the two Nobel laureates discussed truth, beauty, and music. Einstein's step-son-in-law Dmitri Marianoff, who took notes, described Tagore as "the poet with the head of a thinker" and Einstein as "the thinker with the head of a poet," giving the impression that "two planets were engaged in a chat." The Tagore-Einstein dialogues abound with insights into the creativity and the philosophy of the two men and showcase their interest in the arts, especially music. Their first conversation dealt with truth and the nature of reality, with Einstein wondering if truth and beauty existed independently of man. In the second conversation at the Mendel villa, the two talked about family, the German youth movement, and the interplay of chance and predetermination, which led to a discussion about the differences between Western and Indian classical music. Tagore later remembered his host this way: "His shock of white hair, his burning eyes, his warm manner again impressed me with the human character of this man who dealt so abstractly with the laws of geometry and mathematics. ... He seemed to me a man who valued human relationship and he showed toward me a real interest and understanding."

In October 1930, Einstein attended what would be his last Solvay Congress in Brussels. He also stopped to visit "the Royals" in Laeken. They played music, then had dinner together. "I was alone with the Royals for dinner—no servants, vegetarian food, spinach with fried eggs and potatoes, simply that. I liked it there enormously," he wrote to Elsa. After Belgium, he traveled to Zurich, where he received an honorary doctorate from the Eidgenössische Technische Hochschule, his alma mater, and was the guest of honor at the celebrations of the school's seventy-fifth jubilee.

Einstein was scheduled to take a second trip to America in December 1931, this time to California for a stay at the California Institute of Technology. Shortly before he left, Einstein had agreed to write an article on religion for the *New York Times Magazine* titled "Religion and Science." In this and in private letters and other articles published in the period 1927–1932, he outlined his religious beliefs, which were more controversial in the United States than they were in Europe. In this somewhat provocative piece, Einstein professed his belief in a "cosmic religion," which to him was a higher, deeper, much more all-embracing response to life and the universe than were the various forms of organized religion handed down from the distant past. "I assert that the cosmic religious experience is the strongest and noblest driving force behind scientific research," he declared. "The religious geniuses of all ages have been distinguished by this kind of religious feeling, which knows no dogma and no God conceived in Man's image. In my view, it is the most important function of art and science to awaken this feeling and keep it alive in those who are receptive to it." It is a "miraculous order that manifests itself in all of nature as well as in the world of ideas," and the believer "has no use for the religion of fear. A God who rewards and punishes is inconceivable to him." Because "cosmic religion" is on the face of it incompatible with the doctrines of all theistic religions, many Americans concluded that Einstein was an atheist, which he was not. Even Catholic bishop Fulton J. Sheen, whom Einstein admired for his intelligence, quipped, "There is only one fault with his 'cosmic' religion: he put an extra letter in the word—the letter 's'."

Einstein was deeply religious in his own way. As he often proclaimed, he believed in the God envisioned by the seventeenth-century Dutch Jewish philosopher Baruch Spinoza. Spinoza's God "reveals himself in the harmony of all that exists, but not in a God who concerns himself with the fate and actions of human beings," he explained to a Jewish newspaper. To others he wrote that, to him, religion is a humble admiration of a superior spirit that reveals itself in whatever we can understand of the world. In the article "What I Believe," which he wrote for the journal *Forum and Century*, he put forth his personal beliefs and philosophy. In it he wrote, "I have never looked upon ease and happiness as ends in themselves—such an ethical basis I call the ideal of a pigsty. … The ideals which have guided my way … have been Kindness, Beauty, and Truth." He also wondered if God could have created the world any differently and felt that we shall never know the real nature of things.

Chapter 9

ON THE ROAD AGAIN

I do not wish to live in a country where the individual does not enjoy equality before the law and freedom of speech and teaching.

—To the Prussian Academy of Sciences on leaving Germany permanently, April 5, 1933

In December, after they reached the American shore and while their ship was anchored in New York Harbor for five days before it continued its westward voyage, the Einsteins faced the inquisitive press daily. Everyone else wanted a piece of their attention too—public officials, celebrities, and representatives of various organizations. Elsa was in charge of micromanaging the whole ordeal, charging a small fee or asking for a donation whenever her husband posed with someone, gave a short speech, or did other small favors, such as autographing slips of paper or books. The money went directly into a charity kitty. The couple returned exhausted at night to their stateroom.

After a whirlwind of these unexpected activities and the quickly arranged meetings with celebrities, the travelers sailed south into the Caribbean Sea. They stopped in Cuba for two days, then continued farther south, passed through the Panama Canal, and headed northwest along the Mexican coast toward California. In San Diego Harbor, the same fuss accompanied their arrival. From there the Einsteins were whisked off to Pasadena, where they made themselves at home in a cottage near the California Institute of Technology (Caltech) campus. Here in the California sunshine, among the palm trees and citrus groves, they felt as though they were living in paradise.

During his two-month stay in this Eden in early 1931, the outspoken and unflappable Einstein managed to make himself a controversial figure among his conservative hosts. He spoke out against militarism and in favor of pacifism on a number of occasions and advised Caltech students that, as they pursued their technical careers, they must above all show concern for people and their fate "in order that the creations of our minds shall be a blessing and not a curse to mankind. Never forget this in the midst of your diagrams and equations." Among his other scientific activities, he went to Mount Wilson Observatory to meet with astronomer Edwin Hubble. Anxious to have the finest mind in physics among their faculty, Caltech officials offered Einstein a position at an attractive salary anytime he wanted to make a permanent move to California.

Einstein's nonscientific entertainments in southern California naturally included a bit of Hollywood. He was invited to a screening of a film that had been banned in Germany, *All Quiet on the Western Front*, about soldiers fighting and dying in World War I. He was also the guest of actor Charlie Chaplin at a Hollywood premiere of Chaplin's film *City Lights*. The picture of the two charismatic men together at the premiere was circulated widely in the media, and Einstein wrote to his friends back home that Chaplin was as charming in real life as he was on film. He met other celebrities as well, including Helen Keller, social critic Upton Sinclair, and socialist Norman Thomas.

The return trip from California to New York in March was by train, with stops at a Hopi Indian reservation near the Grand Canyon. Here the Hopis, appreciating his pacifist leanings, presented Einstein with a peace pipe. Perhaps being among the first to pun relativity theory, they dubbed him their "Great Relative." A short stop in Chicago followed, where Einstein had just enough time to fire off yet another pacifist speech. Back in New York for less than a full day, he first called pacifists to radical action, then, during a speech at a fund-raiser for Palestine at the Astor Hotel, he urged Jews to cooperate with the Arabs during the settlement process. Then it was back on board ship for the midnight departure and return to Germany, whose political future was now in question even more than when the Einsteins had left home.

Having barely recuperated from his trip, Einstein continued his hectic schedule at home. He delivered two papers to the Prussian Academy of Sciences, then took leave again in May for a month's stay in England. Here he delivered a Rhodes Lecture at Oxford and accepted an honorary degree. After his return that summer of 1931, Einstein sequestered himself at the house in Caputh, determined to continue his pacifist mission. He was convinced that Germany was on the way to becoming an aggressive dictatorship under Adolf Hitler. Ever concerned about war, he wrote countless letters and issued statements on pacifism and the need to refuse military service. In response to a peace demonstration in Flemish Belgium, he wrote about his "militant pacifism," and he called for disarmament and the hope that future generations "will look back on war as an incomprehensible aberration of their forefathers." Disturbed by the political situation and extreme nationalism he saw unfolding around him, Einstein began to consider renouncing the German citizenship he had reacquired on accepting his position at the Kaiser Wilhelm Institute and leaving the country. Hedging his bets about his future in Germany, in the fall he signed a contract with Caltech that would take him to California at least temporarily for the coming winter, giving him more time to consider his future plans.

In the meantime, the world of physics was continuing to see major advances. German physicist Wolfgang Pauli, while in Zurich, predicted the existence of "a little neutral thing," the neutrino, in order to explain where the energy went during beta decay. Austrian logician Kurt Gödel published, at age 25, the paper containing his famous "incompleteness theorem," showing that within any given branch of mathematics there would always be some propositions that cannot be proven either true or false using the rules and axioms of that branch. (Einstein would meet both men again later in Princeton, where Einstein and Gödel would spend many hours walking the streets, discussing physics. Gödel, an eccentric character, feared his food was being poisoned and died after essentially starving himself.) Finally, Ernest Lawrence's "atom smasher," or cyclotron, was put into operation in Berkeley, California.

In December, after short stops in Belgium and Holland, Einstein was once again en route to Pasadena on a ship that took him and Elsa directly to California. They arrived shortly before the end of the year, celebrating New Year's Day 1932 under the palm trees in Pasadena, though signs of the worldwide economic depression were evident in America, too. Throughout his stay, Einstein continued to speak out in favor of pacifism and, often to the embarrassment of those around him, commented on racial discrimination in the United States. In an article written for the official journal of the National Association for the Advancement of Colored People (NAACP), he criticized the rampant racial prejudice he had witnessed. He would later continue to speak out against racism and the lynchings that were still occurring in the South.

During this relatively short stay in California—two months again—Einstein met someone who would change the course of the rest of his life. A prominent educational reformer, fund-raiser, and philanthropist named Abraham Flexner approached him about a new institute he was planning to establish in New Jersey during the following years. Its purpose would be to bring together scholars in the theoretical disciplines, which would require no laboratories. Flexner presented himself as the founding director of an Institute for Advanced Study, which was already established in principle in 1930, to be built through an endowment by two wealthy New Jerseyans. Einstein was vaguely interested in hearing more about the project and agreed to speak with Flexner again in the spring, when both men were planning to be in Oxford. Einstein knew he would eventually leave Germany, but for now he would have preferred to stay in Europe if possible, either in Switzerland or the Netherlands.

In the early spring of 1932, Einstein went to Cambridge to deliver some lectures and then on to Oxford for a visit. Before he left, Flexner paid him a visit as promised, and the two men discussed the new Institute for Advanced Study. When Flexner told him that the Institute would be built in Princeton, where Einstein had lectured and visited 10 years earlier, he became more curious about the possibilities. Flexner offered him an attractive position there—a half-year, annual appointment that would enable Einstein to spend the other half year in Berlin or England or wherever he wished. He agreed tentatively to come to Princeton a year and a half later, in October 1933. In the meantime, Einstein became a grandfather for the second time when Frieda and Hans Albert's baby, Klaus, was born in Zurich.

This year, as a great famine swept the Soviet Union, Jews, mostly Zionists, continued their exodus to Palestine to establish agrarian settlements and begin a new life in their promised land. Germans elected the popular Paul von Hindenburg president for a second time, defeating Hitler by 7 million votes. Hitler, born in Austria, had become a German citizen in February in order to be able to oppose the former general. Though Hindenburg had won the presidency, the Nazis took control of the Reichstag, the German parliament. Hindenburg named someone else chancellor, but, in order to placate the Reichstag, he asked Hitler to become vice-chancellor. But Hitler wanted to be chancellor, nothing less.

Einstein, in the meantime, was not deterred from expressing his political opinions in this dangerous climate. He felt that a Nazi takeover was inevitable. That summer of 1932, he fearlessly continued to speak out against the crisis he saw coming. In one of his last-ditch efforts at advocating peace that year, he took part in a public exchange of letters with Sigmund Freud on the causes of war and the ways to prevent it. The League of Nations' Institute for Intellectual Cooperation published the exchange the following year. In his opinion, "The road to international security lies via the unconditional surrender by nations of some of their sovereignty." He was now beginning to advocate a world government, at least for the purpose of controlling armaments.

In December, Einstein and Elsa once again boarded a ship bound for America for a third stay at Caltech. Because of his "radical" involvements at home, the American consul demanded that, before Einstein could enter the United States, he sign a statement declaring that he was not a member of any subversive organization. Uncharacteristically, Einstein agreed to do so. Before leaving Germany, he told his colleagues and family he would be back in April.

The purging of Jews and Communists from civil service jobs, universities, and organizations in Germany now began systematically. American newspapers and magazines reported the existence of concentration camps in early 1933, after Dachau in southern Bavaria opened its doors in March and imprisoned a group of Communists and other political enemies of the Nazis. The camp quickly gained a reputation for the harsh and sadistic treatment of its political prisoners, who were essentially being used for slave labor. In 1937, the Buchenwald camp was built, mostly for imprisoning German Communists, but Jews were added over the next year as the Nazis' anti-Semitic campaign was set in motion. Homosexuals, gypsies, Jehovah's Witnesses, and some clergy were next.

The Reichstag was burned down at the end of February 1933 while the Einsteins were still in Pasadena. Some say the fire was started by the Nazis, others by the Communists. Possibly both versions are correct. The Nazis are said to have induced a mentally ill Communist sympathizer to set the blaze so that they could then blame the Communists as a whole. Einstein was immediately advised that it was too dangerous to return to Germany. By now he was a wanted man, not so much for being Jewish as for his liberal political views. As part of the Nazi propaganda process, his books were discredited and his character was maligned. When he heard that a price of 20,000 German marks had been put on his head, he joked that he didn't know he was worth that much.

Realizing that it would be impossible to return to Berlin, Albert and Elsa began to make plans. "In view of Hitler, I don't dare step on German soil," he wrote to a friend in late February. The couple wanted to return to Europe though not to Germany. Before leaving Pasadena, Einstein issued statements to the press indicating that he intended to resign from the Prussian Academy of Sciences and called on the world to practice "moral intervention" against the Nazis. "As long as I have any choice, I will only stay in a country where political liberty, tolerance, and equality of all citizens before the law prevail. ... These conditions do not exist in

Germany at the present time," he declared. By now, anti-Einstein feelings were rampant in Germany, and on hearing that he would not be returning to Berlin, the newspaper headlines shouted, "Good News about Einstein—He's Not Coming Back!" He was still careful at this point not to blame all the German people for the Nazi takeover, though even some of his old German friends, such as Max Planck, blamed Einstein for the problems they felt he created for himself. Einstein and Planck, who respected each other tremendously, eventually overcame their differences and remained friends.

The Einsteins left Pasadena by train in March. They stopped in Chicago, then went on to New York. A short detour took them down to Princeton, a university village an hour south of New York, to reinspect what might become their future hometown. In March, they crossed the Atlantic and disembarked in Belgium, where they planned to sort out their options. While still on board the *Belgenland,* Einstein on March 28 wrote a letter of resignation to the Prussian Academy of Sciences in which he showed appreciation for the opportunity to work in Berlin without any professional obligations for 19 years. But, he wrote, under the present circumstances, "dependence on the Prussian government ... is intolerable." In the meantime, the Academy members, for their part, had already read Einstein's statements against the German government while he was still abroad, became indignant, and started proceedings to expel him from their ranks.

The Einsteins set up a temporary residence at Coq sur Mer, a peaceful seaside village surrounded by dunes, and stayed there for half a year. Security guards were sent by the Belgian government to protect them. In April, Einstein's secretary, Helen Dukas, and his valued assistant, Walther Mayer, arrived from Berlin to join them. Margot and Dmitri had already fled to Paris, and Ilse and Rudolf were still in Berlin. During this stay in Belgium, the Einsteins were also guests of Queen Elisabeth and her husband.

In July, Einstein's house, boat, and bank accounts were confiscated by the Nazis "for the benefit of the Prussian state," under laws "concerning the seizure of Communist property and property of enemies of the state." Ilse and Rudolf had quickly tried to save what they could of the contents of the Einsteins' homes in Berlin and Caputh. With the help of the French ambassador, they were able to bring Einstein's library and papers to France by sealed diplomatic pouch. The grand piano and some favorite furniture also made it to safety. From there, the belongings were shipped to America, presumably to New York, then to Princeton. Because Einstein had planned ahead in anticipation of such a crisis, he had kept funds in foreign banks as well and was not in dire financial straits. He took the seizure of his German bank account philosophically, writing to his friend Max Born that he and Elsa would soon have spent it anyway.

Einstein proceeded to resign from all his positions and affiliations in Germany and gave up his German citizenship for the second time. After considering a number of job offers that came his way, he decided to accept Flexner's new proposal of a regular, year-round research position at the Institute for Advanced Study (IAS). The Institute was ideal for Einstein, as if custom-made for him. Then as now, its members had no teaching obligations, and, according to its Web site, it "has no formal curriculum, degree programs, schedule of courses, laboratories, or other experimental facilities. It is committed to exploring the most fundamental areas of knowledge, areas where there is little expectation of immediate outcomes or striking applications; nonetheless, the long-term impact of Institute research has sometimes been dramatic. ... It has no formal links to other educational institutions, but since its founding the Institute has enjoyed close, collaborative ties with Princeton University and other nearby institutions" (www.ias.edu). Furthermore, the faculty is rewarded with handsome wages, prompting some to call the IAS the Institute of Advanced Salaries. Its small campus is idyllic, set on the edge of ancient woods, with a picturesque pond serving as the focal point in its large, well-mowed grassy backyard. It was the kind of place where an independent, freethinking person like Einstein could pursue his goals in peace and in the company of like-minded individuals.

To add to his current troubles, Einstein had learned the previous fall that his younger son, Eduard (Tete), now a 20-year-old medical student, had developed schizophrenia. Einstein was not totally surprised when he got the news, having been aware of Tete's weaknesses and sensitivities since childhood. He felt that, with the state of current medical knowledge, not much could be done for him, that the boy's genes were in charge (the condition was hereditary on Mileva's side of the family) and the affliction would have to run its course. Tete had been placed in a psychiatric hospital in Switzerland, then released when he appeared to be in remission. Soon, however, he was back in the institution. It was in the Burghölzli mental hospital that Einstein saw Tete for the last time in May 1933, on a visit that must have been extremely difficult for him. During this final get-together, he made arrangements to try to ensure Tete's financial security, though Mileva was eventually left with most of the burden in her later years. She devoted herself to caring for Tete at home as much as possible, but his condition continued to worsen until he was permanently hospitalized, with occasional visits home. Mileva's sister, Zorka, meanwhile, who also suffered from schizophrenia, had died in a barn, lying in the hay surrounded by her 43 cats. Mileva continued to visit Tete faithfully until she herself fell ill, suffered a stroke, and died in 1948 at the age of 73. Tete lived into middle age and died in the psychiatric institution in 1965 at the age of 55.

The itinerant Einstein, with Elsa, spent the month of June 1933 in Oxford again. There, he first delivered a Herbert Spencer lecture about the development of the theoretical system—"something occasionally described as mysterious and awe-inspiring"—and the function of pure reason in science. Speaking for the first time in heavily accented English, he maintained that pure thought can grasp reality, and he used mathematical concepts to justify his reasoning. While he was in England, he also met Winston Churchill, whom he found "eminently clever." He noted that Churchill shared his views about the dangers arising in Germany and that the English had made good preparations in the event of war.

He gave another lecture in England, then a third 10 days later in Scotland, where he spoke about general relativity at the University of Glasgow in a George Gibson lecture. The sponsors of the Gibson lecture had asked Einstein to speak about the history of his own scientific work. He agreed to do so because, as he said, it is easier to throw light on one's own work than on someone else's and one should not neglect to do so out of modesty. He traced the work of others who had influenced him and that eventually led to his own discoveries, and he outlined the obstacles he had to overcome in his thinking. In July, he was back in Belgium.

Einstein by now had become disillusioned with pacifism and warned the world repeatedly that Germany was preparing for war. He reluctantly decided that countries that are in grave danger, such as Belgium, had no choice but to depend on their armed forces in the face of a formidable and vicious enemy. Under current conditions, he maintained, he would gladly serve in the military to defend European civilization. At heart, he was still a pacifist, he said, but now Hitler had forced his hand, and he believed that one could be a pacifist only when military dictatorships ceased to exist. Because of his change of opinion, pacifist leaders accused him of being a turncoat to their cause. But he would not budge from his opinion that one had to fight Germany militarily.

In September, the Einsteins returned to England for four weeks before embarking for the United States. Einstein gave one rousing final speech in Europe, at the Royal Albert Hall in London on October 3, warning his listeners about the dangers that lay ahead and praising them for remaining loyal to their democratic traditions. He concluded his remarks with "only through peril and upheaval can nations be brought to further development." The British loved him, considering him, according to an article in the *New Statesman*, a "symbol of the brave and generous outcast, but pure in heart and cheerful in spirit."

Chapter 10

COMING TO AMERICA

> I have warm admiration for American institutes of scientific research. We are unjust in attempting to ascribe the increasing superiority of American research work exclusively to superior wealth. Devotion, patience, a spirit of comradeship, and a talent for cooperation play an important part in its success.
>
> —From an interview, 1921

The small Einstein entourage, without Ilse and Margot and their spouses, who stayed in Paris, boarded their ship on October 10, 1933, to make their Atlantic crossing. After arriving in Princeton on October 17, they spent their first night in town at the large Victorian mansion called the Peacock Inn. A few days later they moved into temporary quarters on the corner of Library and Mercer streets close to the center of town. Their new residence was about halfway between the university's mathematics building, where Einstein would have a temporary office, and the grounds of the Institute for Advanced Study, which had yet to be built. A permanent home on Mercer Street not far from the temporary home would become available a year and a half later. Einstein, at the age of 54, and Elsa, at 57, now began a new life on another continent. Einstein was one of the first faculty members of the Institute, along with mathematicians Oswald Veblen, James Alexander, John von Neumann, and Hermann Weyl. He never again returned to Europe.

In America, a safe haven during those tumultuous years, even during the Great Depression, the Einsteins slowly involved themselves in the intellectual, humanitarian, and political activities that had always been of concern and interest to them. As a new resident in a foreign country, however, Einstein was careful not to appear too outspoken in public, especially on political and Jewish matters. In order to shield him from too many involvements, his new boss, the Institute's director, Abraham Flexner, hovered over him like a mother hen. He even went so far as not to inform him when President Franklin D. Roosevelt invited the Einsteins to the White House, telling Roosevelt that Einstein wished to avoid publicity and live a quiet life. When word about the invitation got around to Einstein through a further letter from the secretary of the Treasury, he angrily confronted Flexner and complained to the Institute's board of trustees, threatening to quit if his life continued to be run by Flexner. Einstein apologized to the Roosevelts. Another invitation to the White House was in the mail by the end of January 1934, and he and Elsa happily accepted it. Even on this stately and memorable occasion, according to his secretary, Einstein did not wear socks. Einstein later told a friend that he was sorry that Roosevelt was so busy as president—otherwise, he would have visited him more often.

Einstein's financial security was now assured. Besides the savings he had set aside in foreign accounts, he received a salary twice that of the average university professor—a real boon in this time of economic uncertainties. His lecture fees were also high. Einstein would have been able to enjoy some luxuries if he had wanted to, but his needs continued to be simple, perhaps even simpler now than when he lived in the more formal German academic environment. Even though his own needs were minimal, Elsa enjoyed the amenities of a middle-class lifestyle. The sacrifices and burdens faced by so many in Europe no doubt caused him to be conscientious about leading an unassuming lifestyle. In time, however, he purchased a small sailboat for excursions on Princeton's picturesque Lake Carnegie so he could participate in the sport he had said demanded the least effort of any sport.

Shying away from politics for the time being, Einstein felt safer applying his energies to humanitarian endeavors. One of his first acts in America in this regard was to give a benefit violin recital in New York as a fund-raiser to help support scientists who were fleeing Germany. But he could not keep quiet for long on what came to be his favorite nonscientific topic: world government. He felt that a centrally run organization in charge of the world's weapons was the best defense against fascism and would ensure world peace.

In May 1934, the family received the bad news that Ilse was deathly ill in Paris. Elsa sailed to Europe to be at her bedside, only to watch her elder daughter die in July at the age of 37. Later that year, Margot and Dmitri came to Princeton, too, while the widower Rudolf Kayser, Ilse's husband, remained in Europe and settled down in the Netherlands. He busied himself by compiling a collection of Einstein's writings titled *Mein Weltbild*, published in America and England as *The World as I See It*. Einstein, in the meantime, spent the summer on the Rhode Island coast with friends, sailing on the Atlantic Ocean.

As politics slipped into the extremes in Germany, with the fascist Nazis on one side and the hard-line Communists on the other, the government was no longer interested in peace, and Germany withdrew from the League of Nations. By this action, it also withdrew from all disarmament talks. After the popular German president Paul von Hindenburg died this year, Hitler finally overthrew the constitutional government and installed himself as "Führer" (leader), demanding the loyalty of all civilians and the military. The salute and cry of "Heil Hitler!" already compulsory among Nazi Party members became a norm in German society. Hitler was granted dictatorial powers over the Third Reich and suppressed all other political parties. Demanding total conformity with his ideas and plans and condemning liberal ideals, he ordered the burning of all books written by Jews and political enemies, including those of Einstein and Sigmund Freud.

To make matters worse, racial hysteria and prejudice drove many Nazis into fanaticism. Any Germans who advocated tolerance were arrested and exterminated—in Hitler's estimation, these were chiefly Jews and Communists. Furthermore, all modernist art was suppressed in favor of realism, and scientific research was being seriously hampered by new regulations. Before 1939, thousands of artists had managed to leave Germany, and hundreds of scientists as well. In neighboring Austria, meanwhile, the Nazis continued to terrorize the Austrians. The Austrian chancellor, Engelbert Dollfuss, a conservative politician who formed an authoritarian government of his own, opposed union *(Anschluss)* with Germany. He paid for his defiance with his life in 1934 when eight Austrian Nazis assassinated him after he had ruthlessly suppressed a national-socialist uprising in the country. The assassins surrendered to government officials and were promptly executed, and the Nazi attempt to take over Austria failed for the time being. In June, Hitler began to remove and assassinate much of the political and military opposition in Germany, and purged the Nazi Party of those he didn't trust. Hearing about an alleged plot by several Nazis against him and his government, he had a number of them executed.

In the mid-1930s, nuclear science, in the meantime, was making strides that would have far-reaching consequences in the next decade. In 1934, the year Marie Curie died, Frédéric and Irène Joliot-Curie produced the first artificial radioactivity by bombarding elements with alpha particles, which were in turn emitted as positively charged helium nuclei from polonium. Intrigued by this work, Enrico Fermi at the University of Rome decided to induce his own artificial radioactivity by using neutrons obtained from radioactive beryllium, but he reduced their speed by first passing them through paraffin. He found the slow-moving neutrons were especially effective in producing emission of radioactive particles and then used this method on uranium. Now he obtained radioactive substances that could not be identified. Fermi was not certain what had occurred. He was unaware that he was on the edge of a world-shaking discovery: he had split the atom, a prelude to a nuclear chain reaction.

Also at this time, Einstein's old friend Leo Szilard had silently filed the first patent application for the idea of a neutron chain reaction. He assigned it to the British admiralty the next year in order to keep the patent a secret. Szilard, who had read H. G. Wells's science fiction novel *The World Set Free*, published in 1914, later acknowledged that he was inspired by the book to discover or invent a nuclear chain reaction. Wells's book prophesied an invention that speeds up the process of radioactive decay that leads to "normal" bombs but ones that continue to explode for days in a chain reaction.

Szilard was an underappreciated scientist who deserves some mention here. A native of Hungary, he was noted for his contributions to nuclear physics, thermodynamics, atomic energy, and molecular biology. He was one of the "Four Hungarians of the Apocalypse," the others being physicists Edward Teller, Eugene Wigner, and Hans Bethe. His ideas and patents, besides the nuclear chain reaction, included the cyclotron, the linear accelerator, and the electron microscope. The adverse economic and political times during which he came up with these ideas prevented him from developing and manufacturing the actual products. Others later expanded on the ideas and received credit for them—even Nobel Prizes. Szilard was also instrumental in drafting the famous "Einstein" letter to President Roosevelt in August 1939 (see later in this chapter).

In November 1934, Einstein was busy with his favorite nonscientific pursuits. He wrote a message on education and peace to be read at a New York conference of the Progressive Education Association. In it, he declared that the United States was in the fortunate position of being able to teach pacifism in the schools: because the nation was not faced with the dangers of a foreign invasion, it was not necessary to inculcate a military spirit in American students. He again advocated an international rather than national military means of defense and a strengthening of international solidarity.

In 1935, the Einsteins purchased a house in Princeton at 112 Mercer Street, less than a mile's walk both to the university and to the future campus of the Institute for Advanced Study, Einstein's professional home until his death. The two-story white clapboard house with the black shutters, wooden front porch, and small front yard has become famous over the years. Visitors to Princeton from all over the world still park their cars across the street and take photographs of the structure, with family members proudly smiling in the foreground to show they have stood where a genius had stood before them. The house has remained a private residence for members of the Institute faculty, and at Einstein's request, it was not turned into a museum or, as he called it, a "shrine."

During his residency in Princeton, Einstein was treated like an ordinary citizen, and his privacy was respected, but many myths and anecdotes about him flourished. He is remembered for being an eccentric, especially in his appearance and in his unconventional behavior for someone of his renown. Carrying an ice cream cone as he shuffled through town, he would stop to greet children and pet dogs, talk to his barber or other familiar figures he saw in town, and walk about in rumpled clothes and an old leather jacket, with his iconic white hair flowing in his wake—a powerful caricature and a cartoonist's dream—and all done without an image consultant. In a letter he kept in his files, a little girl admonished him about his hair: "I saw your picture in the paper. I think you ought to have your hair cut, so you can look better." Even in his most famous portrait, the one by portrait photographer Philippe Halsman that was featured on a postage stamp and on *Time* magazine's "Person of the Century" cover, Einstein wore an old sweatshirt with a pen attached to the collar with a clip. Defending his aversion to socks, he was practical: "When I was young, I found out that the big toe always ends up making a hole. So I stopped wearing socks."

After his death, one could find little physical evidence to prove that the humble Einstein ever lived in Princeton. Until recently, only a street privately owned by the Institute had been named for him, and a small private Einstein "museum" has for many years existed in back of Landau's clothing store on Nassau Street. The Princeton Historical Society, also on Nassau Street, sells a few Einstein-related items, but in 2005 the society's structure will feature an "Einstein room" with memorabilia and furniture donated by the Institute. In April 2005, a statue sculpted and donated by Robert Berks, also the sculptor of the statue of Einstein in front of the National Academy of Sciences in Washington, D.C., was erected in front of Borough Hall. The faculty lounge in Jones Hall, formerly Fine Hall, on the university campus has for years featured Einstein's famous quotation, engraved over the fireplace in German: "Raffiniert ist der Herr Gott, aber boshaft ist er nicht" (The Lord God is subtle, but he is not malicious). It makes no reference to Einstein's residency in Princeton.

For now, however, the Einsteins were interested in becoming U.S. citizens. At that time, applicants for citizenship could file their intention to apply only at a consulate in a foreign country. What could be a better excuse to go to Bermuda than to file for citizenship? With Helen Dukas and Margot in tow, the Einsteins sailed to the island in May 1935. After filing their application, they celebrated at a lavish party given in their honor by the American consul. This trip turned out to be the Einsteins' last excursion outside the United States. They spent the rest of the summer first visiting Massachusetts, where Einstein received an honorary degree from Harvard University, and then vacationing in Old Lyme, Connecticut, a historic seaside village on the Connecticut River. Today, Lyme is better known as the place where, in 1975, Lyme disease was first described.

Meanwhile, in Germany, the government was willfully repudiating the Versailles Treaty that had ended World War I 17 years earlier. It stopped making war reparations to the victors and reintroduced compulsory military service for German men. As Germany began to prosper again, the government passed the Nuremberg Laws. The first, the "Law for the Protection of German Blood and German Honor," prohibited marriages and extramarital intercourse between Jews and Germans (i.e., those of pure German "blood") and the employment in Jewish households of German females under the age of 45. The second law, the "Reich Citizenship Law," stripped Jews of their German citizenship. The Nuremberg Laws in effect formalized the unofficial measures already being taken against Jews. The Nazi leaders stressed that the laws were consistent with the party program, which demanded that Jews be deprived of their rights as citizens.

Hitler's foreign policy and plan for domestic construction were approved by an astounding 98.8 percent of the voters in a new German referendum on March 29, 1936. His influence all over Europe grew as pro-Nazi groups were eager to make their countries part of his Third Reich, which Hitler was determined to expand. Many countries or autonomous regions, hit by the ravages of the worldwide depression, were easy targets for Hitler, who proclaimed that he would now restore the power and prestige Germany lost after its defeat in World War I. Not all were eager to join, however. An Austro-German convention, for one, optimistically reconfirmed that Austria would maintain its status as an independent country, a state of affairs that would change in two years with the *Anschluss* (union of the two countries). During this time, other dictators were in power in Europe, too, and fought resistance from their countrymen: the fascists Francisco

Franco of Spain, Benito Mussolini of Italy, and António Salazar of Portugal and Communist Joseph Stalin in the Soviet Union.

Stalin was not to be outdone by Hitler in his inhumanity. In 1936, he began to conduct his great purge of the Communist Party, arresting, imprisoning, and executing many old-line Bolsheviks and Trotskyist sympathizers, most of them intellectuals and Jews, throughout the land. Less publicized were the purges of younger leaders in partisan, cultural, governmental, and industrial-management affairs. The secret police terrorized the general population as well, with untold numbers of common people punished after spurious accusations. By the time the purges subsided in 1938, millions of former Soviet leaders, officials, and other citizens had been executed or sent to Siberian *gulags* (labor camps). These widespread purges became known as the Great Terror.

In Princeton, Einstein, spurred on by the bad news from around the world, became less reticent about expressing his political views in his adopted country and continued to offer opinions on war and peace. He asserted that every country should be willing to surrender a portion of its sovereignty through international cooperation, that people need to think in international terms, and that, in order to avoid total annihilation, aggression must be sacrificed. His belief was that humans could abolish war through worldwide education, which would lead to an understanding and acceptance of one another's cultures and needs. He also spoke out on education in the schools. At a speech given at a convocation of the State University of New York at Albany in 1936 to celebrate the tercentenary of higher education in America, Einstein emphasized that the aim of education must be to train independently thinking individuals, whose highest life goal should be to serve their community. He urged schools not to use methods of fear, force, and artificial authority—all of which had brought him grief in the German educational system of his youth.

Unfortunately, Elsa did not have a chance to enjoy her new home and furnishings for too long. She became ill with kidney disease in 1936 and had a painful year, some of it spent at Saranac Lake in upstate New York. Some say she never recovered from the death of her daughter Ilse two years earlier, causing a rapid decline in her health. She died just before Christmas, at the age of 60. Einstein did not remain alone in his household, however: he still had the company of his very shy stepdaughter Margot, who was now divorced from Dmitri, and Helen Dukas—and his work. Three years later, his sister, Maja, would join them as well. Einstein claimed that with the death of Elsa, he became more "bearlike," staying in his den and not socializing as he had done with his wife, "who was more attached to other people than I am."

Hans Albert had received a doctorate from the ETH in Zurich, also his father's alma mater, earlier in the year. The following year, he came to the United States to visit his newly widowed father, and in 1938 he resigned his position in Switzerland and also moved to the United States with his family. Late that year, tragedy struck the family again. Hans Albert and Frieda's little son Klaus died suddenly at the age of six, possibly of diphtheria. Einstein wrote to the couple in January 1939, after he had learned of the death, "The deepest sorrow loving parents can experience has come upon you. ... Although I saw [Klaus] only for a short time, he was as close to me as if he had grown up near me." Bernhard was now an only child, but the couple adopted a baby daughter, Evelyn, in 1941. After spending some time in Greenville, South Carolina, working at the U.S. Department of Agriculture, and at Caltech, Hans Albert became a professor of hydraulic engineering at the University of California at Berkeley in 1947. After Frieda died in 1958, he married Elizabeth Roboz. Hans Albert died in 1973 while in Woods Hole, Massachusetts.

Though Einstein continued to give speeches and interviews and carried on a considerable correspondence with friends and others all over the world, he was no longer an active research scientist. His publications were dwindling. He published only one scientific paper in 1937, the year after Elsa's death, and this one was with a young collaborator, Nathan Rosen. In fact, as his working energy declined, his scientific work consisted almost entirely of collaborations with young assistants. One example was the famous article published in 1935 with Boris Podolsky and Rosen, now referred to as the "EPR" paper, in which the three Institute physicists brought to public attention Einstein's critical attitude toward quantum theory. According to their criteria, quantum mechanics did not provide a complete description of physical reality and was an incomplete theory. This paper later created contention among physicists, after the work of John Bell in 1964 propelled it into significance (see chapter 12).

A word about Einstein's English-language capabilities is in order. Einstein came to understand and read English quite well during his time in America, but the middle-aged man never mastered spoken English and often complained about his difficulties with the language in letters to friends. He was able to write and speak simple sentences, and he could read prepared speeches, but in general he reverted to German in long discussions or conversations. Many of his letters and all his papers after he came to Princeton were first written in German, then translated into English by Helen Dukas or his colleagues.

Einstein continued to assist refugees who were fleeing from Germany and sought asylum in the United States, for which they were usually required to have jobs. He tried to find them sponsors or employment, lending or giving money when necessary, sometimes to people he barely knew but whose stories moved him. In 1938, following the annexation of Austria by Germany, a huge number of immigrants sought refuge in the United States. Einstein could no longer handle the flow of appeals that came to him while America was still in a depression and the jobless rate was already high. He realized he was now a part of an ever-growing number of Jewish refugees in America, many of them academics. His closest friends were educated nonacademic professionals such as physicians, artists, and writers with whom he enjoyed speaking about subjects other than physics.

On the political front, President Roosevelt, sensing a tumultuous time ahead in Europe, sent an appeal to Germany and Italy to settle European problems amicably—but to no avail. Late in the year, during the night of November 9–10, 1938, Joseph Goebbels, who was now Hitler's "chief of propaganda and popular enlightenment," launched the infamous *Kristallnacht* (night of broken glass) pogrom in Germany against German Jews. During the night, rampaging mobs throughout Germany and its newly acquired territories freely attacked Jews in the street, in their homes, and at their places of work and worship, killing close to 100 and injuring hundreds more. Destroying almost 7,500 Jewish businesses, they also burned dozens of synagogues, and arrested and sent to concentration camps 30,000 Jews. The Germans maintained that these outbursts were spontaneous, not organized by the government, and that the Jews themselves should be held responsible for the damage done that night. The Nazi leadership seized on the incident to pass laws that would remove Jewish businesses from the economy and "Aryanize" them. Hitler ordered that the "Jewish question" now be coordinated and finally solved. *Kristallnacht* is therefore often considered the beginning of the Holocaust, or Shoah, to give it its Hebrew name, literally "catastrophe."

In the meantime, aviator and medical laboratory assistant Charles Lindbergh had accepted the German Medal of Honor from Hermann Goering, causing outrage in the United States. Following Hitler's appointment as chancellor in 1933, Goering had become chief of the Gestapo, Germany's secret police. Together with two other high-ranking Nazis—Heinrich Himmler, who was the head of the elite, paramilitary "SS" of the Nazi Party, and Reinhard Heydrich, his sidekick in the SS—Goering was responsible for setting up the early concentration camps for political opponents of the Nazis. The following year, Lindbergh made himself even more unpopular at home by criticizing Roosevelt's policies, at which point the president, in return, denounced Lindbergh. The aviator, who lived only a few miles away from Einstein in Hopewell, New Jersey, never returned the medal he received from the Nazis.

Einstein's efforts on behalf of Zionism continued in 1938 with an address given to the National Labor Committee for Palestine in New York City. In "Our Debt to Zionism," he focused on the current troubled times for Jews. Yet, Einstein said optimistically, Zionism renewed a sense of community among the Jews, enabling many of them to escape anti-Semitism and engage in productive work in Palestine.

That year, a huge breakthrough of far-reaching consequences made headlines in physics: Otto Hahn, Lise Meitner, Fritz Strassmann, and Meitner's nephew Otto Frisch, while working in laboratories in Europe, discovered nuclear fission. They were the first to determine that if a uranium atom is bombarded by neutrons, its nucleus would split and cause a chain reaction that would release a tremendous amount of energy in the form of heat and light. When this energy is released all at once, it results in a huge explosion, as in an atom bomb. The experiment was similar to the one done by Fermi in 1934, though Fermi had not realized the implications of his work. This remarkable phenomenon had revolutionary effects not only in physics but soon in world politics as well.

During this time, Einstein's sister, Maja, came to Princeton from Italy. Her husband, Paul Winteler, had been refused entry into the United States because of health problems, which at that time made one ineligible for immigration. He moved to Geneva to live with friends, encouraging Maja to move to Princeton to be in a safer haven with her brother. She hoped to rejoin Paul later, when Europe was more secure. The Einstein household on Mercer Street now consisted of Einstein; Maja; Margot; Helen Dukas; a dog, Chico; and a tomcat, Tiger, named after the Princeton University mascot. Later, a parrot named Bibo, a gift for his seventy-fifth birthday, was added to the household. About Chico he said, "The dog is very smart. He feels sorry for me because I receive so much mail; that's why he tries to bite the mailman."

In May 1939, Einstein addressed a conference on science and religion held at Princeton Theological Seminary. He titled his talk "Our Goal." In it he maintained that scientific and rational means cannot fully serve to influence a person's convictions and beliefs—that they have their limits. Therefore, the most important function of religion in our social lives is to make clear a society's values and goals—that is, the powerful traditions that have given a foundation to aspirations and values. Because they have already worked well in a healthy society and thus proven themselves valuable, traditions do not need to be justified.

By now, Einstein was becoming an outsider in the world of physics, no longer involved with mainstream research. He continued to criticize and debate quantum mechanics and to search for a unified field theory, leaving physicists who did not share his views bemused if not annoyed. Unable to achieve breakthroughs of his own anymore, his scientific satisfaction came vicariously through the accomplishments of his younger associates.

Nevertheless, because of his reputation as a "genius" and humanitarian, Einstein's name continued to carry weight, and his involvement in a project often led to its success. Therefore, for example, during the summer on a muggy July day in 1939, his old friend Leo Szilard came to pay a visit at Einstein's summer retreat at Peconic Bay on Long Island. Another physicist, Hungarian-born Eugene Wigner, was at his side. They informed Einstein about the new results obtained with uranium-235, that is, about nuclear fission. The visitors feared that if German scientists were aware of the possibilities inherent in using such tremendous energy, they might begin work on an atom bomb. They reported that Frédéric Joliot of France, for example, had immediately recognized the implications of the discovery made independently by him and his wife and by Enrico Fermi in 1934. Joliot had ordered six tons of uranium oxide from the Belgian Congo as well as heavy water from Norway, both of which are vital ingredients in making a successful nuclear device. Heavy water is chemically the same as regular water except the two hydrogen atoms are replaced with deuterium atoms. Deuterium has one extra neutron, which makes heavy water about 10 percent heavier than regular water. This kind of water is used as a moderator and coolant in nuclear reactors. A moderator slows down the neutrons that are emitted in a chain reaction, increasing the fission reaction rate and enabling a sustained chain reaction. A chain reaction is a process in which neutrons released in fission (splitting or breaking up into parts) produce an additional fission in at least one further nucleus. This nucleus in turn produces neutrons, and the process repeats. The process may be controlled (for use in nuclear power) or uncontrolled (for use in nuclear weapons). The materials were shipped to England for safekeeping in case of a German invasion.

Einstein, highly disturbed by the news, agreed to get involved. By August, Szilard had drafted a letter to Roosevelt, warning the president about the military uses of atomic energy. The letter stressed that a catastrophe might lie ahead if Germany succeeded in building a bomb and the United States neglected to do so. Though Einstein signed the letter immediately—his was the only signature on the letter—it was not delivered to Roosevelt until October, more than a month after World War II had already begun in Europe. The United States had not yet entered the war, but Roosevelt, now eager to prepare America for possible future involvement, thanked Einstein for the important news. He proceeded to appoint a committee to do a year's research on uranium and to study the possibilities of nuclear energy. This decision led to the organization of the top-secret Manhattan Project at Los Alamos in the New Mexico desert and the designing and assembling of the atom bombs detonated in 1945.

On September 1, 1939, Germany invaded Poland and annexed the northern German-speaking port city of Danzig (Gdansk). Within two days, Britain and France declared war on Germany while the United States remained neutral. Germany continued to overrun western Poland and took Warsaw, the capital city, immediately installing a Nazi governor-general. At the same time, the Soviet Union invaded Poland from the east. In England, meanwhile, large numbers of children were evacuated from London, and those without relatives in safer parts of the country to the west were billeted with kindhearted strangers. Some mothers took their children to Canada, but, with U-boats operating in the Atlantic, not all survived the crossing. In the United States, the economy began to recover as the European nations frantically ordered arms and materiel in preparation for war.

Even though war was being waged overseas, preparations were made for a World's Fair to be held in New York in 1939–1940. Einstein and the members of his household went to Flushing Meadows to enjoy the exhibits. The fair's theme was "The World of Tomorrow," and its mission was to promote science and technology as the means by which to achieve economic prosperity and personal freedom. Coming right between the Great Depression and America's participation in World War II, the fair symbolized hope for the future. On request, Einstein submitted a statement, "To Posterity," for the fair's time capsule. In it, he recalled some of the world's technological achievements, then informed future generations that people were presently living in both economic fear and fear of war.

At the beginning of the new decade, as Einstein became an American citizen and took his oath of allegiance in Trenton, New Jersey, the war was spreading in Europe and also in the Far East. Under skies heavy with military planes, Winston Churchill became prime minister of Britain after Neville Chamberlain resigned. He passionately delivered his famous "blood, sweat, and tears" speech to rally his countrymen and prepare them for what lay ahead. In the meantime, Germany invaded Norway, Denmark, Holland, Belgium, and Luxembourg; the Dutch and Belgians surrendered, and German occupation began in the two countries. Italy declared war on France and Britain, and the German army entered Paris in mid-June. Crossing the English Channel in large numbers, the Luftwaffe, Germany's air force, proceeded to conduct air raids over London, at first both day and night, then only at night because too many planes were shot down during the day. All over Europe, cities were being evacuated as their citizens tried to find safety from air strikes for their families, often to no avail. In the Far East, Japan, China, Burma, and the Philippines clashed. Japan, Germany, and Italy—the Axis powers—signed a military and economic pact to help one another achieve their goals. Throughout all of this turmoil and disaster, many of Europe's greatest scientists, most of them Jews, managed to flee to the United States and Canada.

In America, voters elected the popular Roosevelt as their president for an unprecedented third term. In anticipation of war, Roosevelt asked Congress for a huge defense budget and urged the production of 50,000 airplanes. At the same time, Congress created the Selective Service System—"the draft"—for compulsory military service, and all men between the ages of 21 and 36 were required to register immediately. The Smith Act, requiring all aliens to register, was quickly enacted.

Only one month before Japan's attack on Pearl Harbor, the U.S. government finally set into motion the plans for a massive technological and scientific enterprise in support of the development of atomic energy. Government officials created the highly classified Manhattan Project, to be carried out secretly in laboratories by skilled physicists and technical experts at Los Alamos. Early in 1942, General Leslie Groves was appointed director of the whole Manhattan Project. He in turn appointed physicist J. Robert Oppenheimer as director of research at the Los Alamos National Laboratory. Oppenheimer's responsibility: oversee the development of the atom bomb.

When the Manhattan Project was established, its first major challenge was to find acceptable and plentiful sources of fuel for the bombs. Both uranium-235 and plutonium-239 were likely candidates; they were the only two fuels known at the time that could sustain the chain reaction required to set off an atom bomb. The plutonium or uranium in the bomb undergoes fission in an arrangement that generates enormous energy and has great destructive potential. But there were no laboratories where these fuels could be produced, so they had to be built—one for uranium, the other for plutonium. After relocating more than 1,000 rural families in Tennessee to build one of the laboratories, government employees began work at the Clinton Engineer Works (in 1943, the name was changed to Oak Ridge) to extract uranium-235 from natural uranium ore, uranium-238. By 1945, scientists were able to produce enough bomb-grade uranium-235 for use in Los Alamos. To produce plutonium, which does not occur naturally, the government built a 500,000-acre facility starting in 1943 in the state of Washington along the Columbia River; it became known as the Hanford Engineer Works. In the process, 1,500 families were relocated, and three small towns, farms, and vineyards disappeared forever.

By now, the United States was already at war. After the U.S. ambassador to Japan had warned Roosevelt of a possible Japanese attack on the United States, Pearl Harbor was hit on December 7, 1941. The United States and Britain declared war on Japan almost immediately. On December 10, Germany declared war on the United States. Italy followed suit, and the United States declared war on both these fascist European countries.

Already during this time, Italian physicist and Nobel laureate Enrico Fermi, who had fled fascist Italy with his Jewish wife to come to the United States, and his colleagues were at work at the University of Chicago to produce a controlled nuclear reaction. On the afternoon of December 2, 1942, a year after Pearl Harbor, they witnessed the first controlled energy release from the nucleus of an atom in their small underground laboratory. Now the laboratories at Oak Ridge and Hanford would be able to develop ways of obtaining nuclear fuel for atomic weapons.

Einstein was also ready to help in the defense effort, but he was unable to do so: the government would not give him the required security clearance. His newly adopted homeland considered the new citizen a security risk because of his controversial political opinions, and his fearlessness in associating socially with leftists had made him a suspect American. FBI agents had been following his every move for years and continued to collect information even after his death. They amassed a file of more than 1,400 pages that can now be viewed on the FBI's Web site. In its prefatory statement, the site claims that Einstein was a Communist, even though within the files the FBI had concluded that Einstein did not appear to have Communist leanings, did not belong to Communist organizations, and was not involved in subversive activities. Had FBI agents known that Einstein's warnings helped establish the Manhattan Project to defend freedom and the nation, they might have been more likely to consider him a loyal American. In later years, he proclaimed, "I have never been a Communist. But if I were, I would not be ashamed of it." He felt that Americans were hysterical about the dangers of Communism, noting that western European countries had no such fears even though Communist parties were in existence there.

Even though the U.S. government would not give Einstein the security clearance necessary to work on the Manhattan Project—it is not clear whether he would have agreed to do so anyway—he showed his patriotism in other ways. He became a consultant on explosives for the U.S. Navy, for which he was paid $25 a day. A colleague joked that Einstein was now a member of the navy but wasn't required to submit to the short haircut. Visitors also came up from Washington seeking information on, for example, the optimal detonation of torpedoes. He felt useful in being part of the effort to defeat fascism. Unbeknownst to him at this time, he was spared from any direct involvement in the development of the most destructive weapon ever built.

A further contribution to the war effort in 1944 enriched the American war chest considerably. The Book and Authors War Bond Committee, which purchased war bonds with money made by auctioning important manuscripts or manuscripts written by famous people, asked Einstein if he could donate his 1905 manuscript on relativity. The original was no longer available, but the physicist agreed to copy it by hand from the printed published version. His secretary, Helen Dukas, read it out loud as he wrote down his words from so long ago. Occasionally he stopped, saying that he should have written some specific passage much more simply or differently. He contributed another manuscript for the same cause. Both manuscripts were auctioned in February: the relativity copy brought $6.5 million and the other manuscript $5 million, huge sums in those days and even now. The anonymous buyers donated both copies to the Library of Congress.

Now, in the midst of war, Einstein felt as anti-German as ever, especially as he heard stories such as that of the Warsaw Ghetto uprising in Poland. The Nazis, who had taken control of Poland, had restricted Warsaw's Jewish population to a ghetto. According to the U.S. Holocaust Memorial Museum's Web site, in the summer of 1942 the deportation of more than 300,000 Jews from the ghetto began, with 250,000 going to the concentration camp at Treblinka. News of the exterminations that took place there soon came back to the ghetto's approximately 56,000 remaining residents. The following year, a cluster of mostly young people in the ghetto formed a clandestine group that urged the next targeted victims to resist going to the railroad cars that would take them to their deaths. When the German soldiers came to carry out their orders, the group fired on them with weapons that had been smuggled into the ghetto, and the troops retreated. A few months later, the soldiers were back again in larger numbers to remove the remaining inhabitants. The young Jewish fighters fought back bravely for a month. In the end, however, they lost the struggle as the Germans crushed the resistance, set fire to the ghetto, and captured virtually all the residents, killing 7,000 of them on the spot and deporting the remainder to forced-labor and concentration camps. After hearing about the heroism in Warsaw, Einstein wrote a famous two-paragraph statement paying tribute to the victims. He famously blames "the Germans as an entire people" for the mass murders in Europe because they elected Hitler by a large majority even though they were made aware of his long-term intentions in his book and speeches.

Einstein continued to be concerned about the arms race and wrote President Roosevelt another letter in March 1945 at the request of Leo Szilard. He asked that the president speak with Szilard about the concern of many physicists that there was a lack of contact between those conducting classified research and those making policy. Einstein did not know the exact nature of the Manhattan Project, but presumably he was told that it could have grave consequences. Roosevelt, however, died on April 12 and never saw the letter. He also did not live to see the other major events occurring over the next month: the end of German resistance, Hitler's suicide on April 30, and the official German surrender on May 7. The Allied offensive finally emerged victorious in Europe after a hard-won victory, and VE (Victory in Europe) Day was declared on May 8. But Japan had not yet surrendered in the Pacific.

In 1945, before the end of the war with Japan, the U.S. government issued a document called the Smyth Report. Among other things, it disclosed that Einstein had made Roosevelt aware in 1939 that the Germans might have the ability to build a bomb and that he was indirectly responsible for the atomic research conducted by the United States. The report, written at the end of June 1945, summarized the atom bomb project and reported that scientists and the military were expecting the bomb to be tested imminently—that "a weapon has been developed that is potentially destructive beyond the wildest nightmares of the imagination." A brighter note, according to the report, is that peaceful uses of atomic energy were possible as well. Most experts agree that the United States would have built the bomb even without Einstein's communication, as the technology to do so was already present both in the United States and in England.

On July 16, scientists at Los Alamos were ready to test an atom bomb at Alamogordo. The test was successful, and the scientists involved in the project were elated. Many, however, became reflective and alarmed about their achievement and what it would mean in terms of world peace. On hearing that the bomb project was completed, the guilt-ridden Leo Szilard quickly circulated a petition, to no avail, asking the United States not to use the bomb against Japan on moral grounds.

By now, most of us have seen film of the enormous mushroom cloud rising over the test area in the desert. Three weeks later, the explosion was no longer a test. On August 6, the U.S. B-29 bomber *Enola Gay* left the small South Pacific island of Tinian, roared into the night sky with a bomb and twelve men, flew some 1,500 miles toward Japan, and dropped a uranium-235 bomb, nicknamed "Little Boy," on Hiroshima. Three days later, the bomber *Bock's Car* dropped a plutonium-239 bomb, "Fat Man," on Nagasaki. The shock waves, fire, and radiation claimed about 70,000 lives in Hiroshima and 40,000 in Nagasaki instantly, and thousands more died later. Japan surrendered, and World War II ended on August 14, 1945. Close to 50 million people had lost their lives worldwide by the end of the war, including soldiers, civilians, and those who had died in concentration camps.

In later years and in hindsight, after witnessing the devastation and agony caused by atomic weapons, Einstein continued to be troubled by the most devastating end product of his discoveries of 1905. If he had known that the Germans would not be successful in building a bomb, he said, he would never have signed the letter to Roosevelt. Though German scientists were aware of the possibilities of nuclear chain reactions, they did not build a bomb because the resources were not available to them, and even if they were, the required procedures could not have been carried out there under war conditions.

In his new crusade to ban the bomb, in a magazine article of November 1945, Einstein advocated a world government whose purpose would be to control use of the atom bomb and all armaments. It should be founded by the three great military powers: the United States, the Soviet Union, and Great Britain. This world government would have power over all military matters and be able to intercede in countries where oppression occurs. Even a tyrannical world government is preferable to the far greater evil of wars, he wrote. Any kind of national government has a certain amount of inherent evil, and another major war was even more evil than that. Later, the Russian Academy of Science denounced Einstein for advocating such a system.

Coincident with the end of the war was Einstein's official retirement from the Institute for Advanced Study. He now began to resume his pacifist activities by continuing to advocate a world government and writing on war and peace and disarmament. "The release of atomic energy has not created a new problem," he wrote. "It has merely made more urgent the necessity of solving an existing one." The existing problems were much the same as always. Much earlier, Baroness Bertha von Suttner, who had won the Nobel Peace Prize in 1905, had remarked that improving the laws of war was like regulating the temperature while boiling someone in oil. There could be no "improvements" to war. Einstein now felt that it was the moral duty of scientists engaged in basic research not to cooperate in military matters, warning, "I do not know how the Third World War will be fought, but I can tell you what they will use in the Fourth—rocks!"

Chapter 11

FINAL YEARS OF AN ENFANT TERRIBLE

> I have become a kind of *enfant terrible* in my new homeland because of my inability to keep silent and swallow everything that happens here.
>
> —To Queen Elisabeth of Belgium, March 28, 1954, regarding his outspokenness during the McCarthy era

At a dinner in New York at the Astor Hotel honoring Nobel Prize winners in December 1945, Einstein voiced his concerns about the atomic weapon his fellow physicists had delivered to the world. He soberly uttered the words: "The war is won, but the peace is not." He had in mind the continuing arms race among nations.

In the postwar years, Einstein became increasingly disturbed about the atom bomb and its possible use. He returned to his prewar pacifism by speaking out against militarism and in favor of a world government, sometimes advocating "Gandhi's methods" to achieve it. He believed that as long as individual nations maintained separate armaments, there would always be wars. He chaired the newly formed Emergency Committee of Atomic Scientists, which concerned itself with the peaceful uses of atomic energy. From now until the end of his life, he immersed himself in the issues of peace and world government. As to his relationship with Germany, he would not reconcile with the country and wanted nothing to do with any Germans except for the few who had been steadfast in their opposition to Hitler. He scorned German Jews such as philosopher Martin Buber who returned to Germany to accept awards, and he continued to rebuff the efforts of Germans to reconcile with him. He declined a foreign membership in the prestigious Max Planck Institute and the honorary citizenship of Ulm, the city of his birth. He would do the same in the case of West Berlin in 1952 and declined membership in the German section of the International Organization of Opponents of Military Service in 1953.

In 1947, in his campaign to settle disputes through nonaggression, he accused America of placing the importance of military power above all other factors that affect relations among nations, resulting in a hawkish mentality in government. He warned that Germany's similar attitude, beginning with Otto von Bismarck and Kaiser Wilhelm II, resulted in Germany's decline in less than 100 years. Now, he argued, a military mentality is even more dangerous because of the more powerful weapons. He also wrote in favor of enacting international laws prohibiting the building and use of the atom bomb.

After many years of illness and hardship, Einstein's first wife, Mileva, died at the age of 72 in Zurich in 1948. This left Eduard without the attention and love his mother had provided over the years. Carl Seelig in Zurich, Einstein's biographer and a man he had grown to respect, had taken an interest in Eduard and now visited him regularly. Einstein was grateful for Seelig's attentions and apologized for his own negligence of Eduard. He wrote Seelig that he did not understand his own reticent behavior toward his son and his reluctance to be in touch with him, even through letters. "I believe I would arouse painful feelings of various kinds in him if I were to make an appearance in whatever form," he rationalized.

The creation of the new state of Israel had been under way since 1947. The United Nations voted to partition Palestine into two sovereign states, one Arab and one Jewish, and the British would give up their mandate over Palestine. The 1.3 million Arabs rejected the plan to give up half their land to the 600,000 Jews. In 1948, a provisional Israeli government was announced by Jewish leaders, and the British withdrew within two months and Israel declared its independence. The Arabs did not give up without a fight, and the Arab states surrounding the new nation invaded, but the Israelis drove them back. Around 750,000 Arabs fled or were sent to refugee camps. About 800,000 Jews fled the Arab nations, with half a million going directly to Israel, which had opened its borders to all Jews. Within a year, Israel signed armistices with the Arab states, and a precarious peace was declared in the area. Chaim Weizmann became president and David Ben-Gurion prime minister of the first Jewish state in nearly 2,000 years.

After Weizmann died in 1952, Einstein was offered the presidency of the new nation. He declined, saying he was deeply honored by the gesture but not suited to the job. Moreover, he had conflicting feelings about the new state, whose policies he didn't always support. Israeli leaders, knowing that Einstein's skills lay elsewhere, were relieved that he turned down the offer. Prime Minister Ben-Gurion pleaded with his aide Yitzhak Navon, who later became president of Israel himself, "Tell me what to do if he says yes. I had to offer the post to him because it is impossible not to. But if he accepts, we're in for trouble."

In the United States, meanwhile, the Marshall Plan, also known as the European Recovery Program, was instituted to help European economic recovery after World War II. Secretary of State George C. Marshall had proposed the plan in a commencement speech he delivered at Harvard University in 1947. He outlined a sensible and sound policy by which the United States could provide aid to reduce hunger, homelessness, sickness, unemployment, and political restlessness among the 270 million people in 16 nations in western Europe who had suffered the effects of six years of war. The economies of the regions were devastated, millions were homeless, and the destruction of agriculture had led to conditions in which many on the continent were near starvation. Many of the greatest cities of Europe were in ruins, and others were severely damaged. The transportation industry had been especially affected because railways, bridges, and roads had been heavily bombed and many merchant ships sunk. Yet Europe had spent its money fighting the war and had no financial resources for reparations. The United States, on the other hand, was in better shape because its land had not been destroyed and it had suffered only one attack, in Hawaii.

Marshall Plan funds of about $13 billion (around $100 billion in today's money) were directed toward strengthening the economic superstructure of Europe over a period of four years. Its goal was made easier by the fact that the European people were well educated, industrious, and able to rebuild their own lands, if only with financial aid. The plan worked so well at reducing postwar suffering relatively quickly that George Marshall was awarded the Nobel Peace Prize in 1953. He had already been named *Time* magazine's "Man of the Year" in 1948.

Besides advocating pacifism and crusading for a world government, Einstein maintained his interest in religion. In 1948, for example, he published an article in the *Christian Register* on the compatibility of religion and science. He conceded that the issue is complicated because people can agree on what science is, but they often differ on their definition of religion. The part of religion that is most likely to cause conflict is its mythical aspect—which is only one facet of religion—and myths are not necessary for the pursuit of religious aims. The only organized religious group he seemed to admire was the Quakers, or Society of Friends: "I consider the Society of Friends the religious community that has the highest moral standards. As far as I know, they have never made evil compromises and are always guided by their conscience. In international life especially, their influence seems to me very beneficial and effective," he wrote to an Australian correspondent. And to a nun, he wrote, "A man's moral worth is not measured by what his religious beliefs are, but rather by what emotional impulses he has received from Nature during his lifetime."

Einstein's sister, Maja, had planned to return to Europe after the war to rejoin her sick husband, but it was not to be. After several years of contented living in her brother's home, she suffered a stroke and remained bedridden. Einstein wasn't doing too well, either, continuing to suffer from his gastrointestinal problems. A few months before he turned 70, doctors thought he had an intestinal ulcer and admitted him to Jewish Hospital in Brooklyn. The diagnosis, however, was an aneurysm in the abdominal aorta. The main aorta in his stomach had become dilated and blood-filled to the size of a grapefruit, and it was in danger of bursting once the aortic wall weakened. Surgery was too risky at the time, so Einstein went home after a few weeks. Afterward, he recuperated in Florida for a while and worked on his "Autobiographical Notes" for inclusion in a volume on living philosophers. He referred to these notes as his "obituary."

After his return to Princeton, Einstein found that Maja's condition had worsened, and, as the faithful big brother, he devotedly read to her every night. In middle and old age, the siblings had grown to resemble each other, especially in their hairstyle, and from the back it was difficult to tell one from the other. A friend, Lily Kahler, saw them as "two old people sitting together with their bushy hair, in complete agreement, understanding, and love." Maja suffered for several years and died in Princeton in 1951 at the age of 70. Maja's departure left a big hole in Einstein's life. Her husband, Paul, died the following year.

Einstein became a septuagenarian in 1949, and the event called for a celebration. Three hundred scientists assembled in Princeton to pay tribute to him and to hold a symposium on his contributions. When Einstein entered the auditorium for the festivities, the audience turned around and fell silent as he slowly walked down the aisle. Then they stood up in a rousing ovation.

The following year, Einstein, becoming ever more aware of his ill health and age, signed and sealed his last will. He named his friend Otto Nathan, an economist, as executor and, along with Helen Dukas, a trustee of his literary estate. Overlooking his earlier disputes with the institution, he willed his papers to the Hebrew University of Jerusalem, to be sent there after the death of Nathan and Dukas. His violin was to go to his grandson,

Bernhard, and his financial assets to Helen Dukas, his sons Hans Albert and Eduard, and his surviving stepdaughter, Margot.

In February 1950, President Truman had some astounding news: the United States had successfully built a hydrogen bomb—a bomb even more powerful than the atom bomb. Hungarian-born physicist Edward Teller was named its "father." On hearing the news, the enraged Einstein decided to speak on a nationwide television broadcast filmed in Princeton. If the bomb were ever used, he cautioned, the planet would be poisoned by radioactivity and all life on Earth would be annihilated. In June, in a United Nations Radio interview, he again warned, "Competitive armament is not a way to prevent war. Every step in this direction brings us nearer to catastrophe. ... I repeat, armament is no protection against war, but leads inevitably *to* war. ... Striving for peace and preparing for war are incompatible with each other."

Around this time, many Americans became alarmed by a political episode that blighted the nation. The end of the war, after the Soviet Union ceased being a U.S. ally, had brought increased fear of Communism in the United States, a fear that was driven mostly by conservative politicians. For several years, the House Un-American Activities Committee (HUAC) had already existed in the House of Representatives, and after the war the committee initiated investigations into Communist influence in Hollywood. In September 1947, HUAC subpoenaed 41 people to testify at formal hearings. Nineteen of these testifiers were considered "unfriendly" because they refused to cooperate with the committee. Eleven eventually came but refused to answer questions, denounced the committee, and were therefore held in contempt of Congress and sentenced to brief prison terms. After their release, they were blacklisted from the movie industry for many years. The blacklist itself was not developed by HUAC but by a group of studio executives who agreed not to rehire them. The "friendly" witnesses, including actors Gary Cooper, Ronald Reagan, and Robert Taylor, agreed to testify about Communist infiltration in the Hollywood movie industry. HUAC conducted another round of Hollywood investigations in 1951.

And now enter Senator Joseph McCarthy, a junior senator from Wisconsin. McCarthy had decided to dedicate himself to promoting the Red Scare and going on Communist witch hunts, in which he participated from 1950 to 1953 as chairman of the Senate's Permanent Subcommittee on Investigations. McCarthy was not after Hollywood but after government infiltrators. In a speech in 1950, he dramatically waved a sheet of paper that he claimed contained the names of Communists and Communist sympathizers who had infiltrated the State Department. A special

Senate committee investigated the charges and found them groundless. Unfazed, McCarthy used his position to continue to wage a hysterical anti-Communist crusade. With little if any proof of his charges, he denounced civil servants and government officials he considered politically suspicious. He held a series of highly publicized hearings in front of Congress in which individuals were required to defend themselves. The committee's zeal, with little or no evidence to back up its claims, often destroyed people's lives and offered no apologies, leaving the victims to fend for themselves.

Einstein advised many of the targets not to cooperate in these hearings, including those of the Internal Security Subcommittee (nicknamed "Sissy"), the Senate's equivalent of HUAC. Viewing the process as a violation of an individual's civil rights, Einstein, in a letter in May 1953, advised teacher William Frauenglass of Brooklyn, "I can see only the revolutionary way of non-cooperation in the sense of Gandhi's methods. ... Refusal to testify must be based on the assertion that it is shameful for a blameless citizen to submit to such an inquisition and that this kind of inquisition violates the spirit of the Constitution." To another he wrote that citizens in a free country are not obligated to give an accounting of their party memberships.

In 1953, McCarthy decided to take on the U.S. Army. He claimed that this branch of the military was infiltrated with Communists as well, and ordered investigations to proceed. During hearings, the army's attorney felt McCarthy had gone too far in his accusations and angrily exclaimed, "Have you no sense of decency, sir? At long last, have you left no sense of decency?" McCarthy's influence now began to wane, especially when televised hearings in 1954 allowed millions to view his methods for the first time. By the end of the year, the Senate, in a resolution, voted to censure the senator for "conduct that tends to bring the Senate into dishonor and disrepute." Since then, "McCarthyism" has come to denote a witch hunt carried out by governments to seek out and punish unapproved thought or political opinion. McCarthy died two and a half years later of hepatitis, a result of alcoholism.

Disillusioned with politicians, Einstein advocated keeping them out of science policymaking. In 1952, he wrote that politicians should neither control the sciences nor impede free scientific exchange with other countries. He was distrustful of politicians, maintaining they were determined, even in peacetime, to organize our lives as if we were at war, promoting work geared toward winning a war.

Einstein continued to become less physically active in 1953. He felt his age as his health deteriorated. His strolls down the streets of downtown

Princeton that summer became rare, and as he listened to the whirring sounds of the 17-year cicadas, he must have known this would be the last time he would witness this spectacle. The halcyon days of eating ice cream cones, petting dogs, and chatting with neighbors were coming to an end. He managed to write a short paper on the unified field theory, his final one on the subject. With some optimism he said he was certain about the mathematical concepts but uncertain about the physical ones.

In 1954, Einstein, though he had developed hemolytic anemia, a condition in which the bone marrow is unable to compensate for premature destruction of red blood cells, worked on his last scientific paper. His collaborator was Bruria Kaufmann, a young physicist who was born in Palestine but educated in the United States. Throughout the year, she visited him at his home on Mercer Street and reported on progress in calculations, which were often wrong and had to be redone. They finally considered their paper complete and published it the following year.

At the height of the McCarthy and HUAC hearings in 1954, the U.S. government alleged that former Los Alamos director J. Robert Oppenheimer, the "father of the atom bomb," had Communist sympathies. The fact that his brother, Frank, had been a member of the Communist Party in the 1930s made him even more suspect. The government's Personnel Security Board conducted hearings, found him unpredictable and untrustworthy, and withdrew his security clearance. They also stripped him of his position on the Atomic Energy Commission, ending Oppenheimer's influence on scientific policy. The committee based its decision partly on critical testimony from Edward Teller, the "father of the hydrogen bomb," whose construction Oppenheimer had opposed. (Coincidentally, a second hydrogen bomb was exploded on Bikini Atoll in the Marshall Islands that year, vaporizing three of the islands.) Einstein had supported Oppenheimer, who had been director of the Institute for Advanced Study since 1947, both in his opposition to the hydrogen bomb and during his dealings with the investigating committee. A hero of the left and liberals, Einstein now became an antihero of the right, which wanted him stripped of his citizenship and deported. The specter of intellectuals being persecuted and civil liberties and freedom of expression being denied was painful for the aging scientist, who was all too familiar with this situation.

Einstein's pleasures this late in his life had become few. He shied away from celebrations such as the one on his seventy-fifth birthday on March 14, 1954, after which he complained he ate too much birthday cake. He received countless letters and gifts from people he didn't know. One gift was a parrot he named Bibo, who arrived at 112 Mercer Street like an ordinary piece of mail. Besides answering his birthday mail, Einstein spent the next few weeks trying to cheer up the anxious and depressed bird with jokes. At other times, he would venture out to sail with a friend or two in his boat on Princeton's tranquil Lake Carnegie. By now, his doctors had forbidden him to smoke his pipe, but he still liked to hold it and chew on its mouthpiece. The violin playing had stopped long ago, being too physically demanding for him in old age; but he still enjoyed sitting down at the piano or listening to music on the radio or records. He also continued to receive an astonishing number of guests, many of them coming to him with urgent requests, some out-of-towners arriving at his house even in the middle of the night.

As we know, Einstein was a widower for almost 20 years. Women often wrote to him, detailing their special attributes and homemaking skills and offering to become his wife. Others pursued him closer to home, and he often returned the favor. One of his favorite "best friends" while in America appears to have been Margarita Konenkova, who was married to a well-known sculptor who made a bronze bust of Einstein that is currently in the Institute's library. Margarita was later rumored to be a Soviet spy. She and her husband were Soviet émigrés who lived in Greenwich Village in New York, about an hour from Princeton, from the early 1920s until 1945, when they were recalled to the Soviet Union. Einstein did not seem to know that she was suspected of being a spy. Even if she were, he had no access to any classified information that the Soviets might have wanted. After Konenkova's departure, Einstein befriended Johanna Fantova, who remained his lady friend until his death. Fantova, who was about 20 years younger, stood by him through his physical decline. During the last year and a half of Einstein's life, she kept a journal of her telephone conversations with him. Judging by the journal's entries, Einstein still discussed the topics that had always been of interest to him, in addition to the health problems one can expect to hear about at that age.

In October 1954, Einstein wrote a famous letter in which he claimed that if he could live his life again, he would not become a scientist. He said he would rather become a plumber or a peddler, "in the hope of finding that modest degree of independence still available under present circumstances." By this he did not mean that he would give up doing physics; he meant he would be less encumbered by the other obligations that being a researcher and academic entail. Perhaps he was becoming nostalgic about his days in the Patent Office in Bern when he was not yet famous and had more freedom, ability, and time to think creatively about the subjects and ideas that attracted him.

Early in 1955, philosopher and writer Bertrand Russell approached Einstein, asking him to issue a joint statement with other scientists who opposed the international arms race. Russell, who had won the Nobel Prize in literature in 1950, had drafted a document denouncing the possibility of a nuclear war in the wake of the Cold War that he wanted to submit to the U.S. Congress. There would be no winners or losers in a war in which a hydrogen bomb was detonated, they said, only a permanent state of catastrophe, and they appealed to Congress as "human beings to human beings." Russell sent the statement to Einstein, who signed it on April 11 and returned it to Russell. This statement, which became known as the Russell-Einstein Manifesto, was signed in alphabetical order by nine other prominent scientists and was subsequently endorsed by thousands of others worldwide.

That same day or the next, Einstein started to feel gravely ill, feeling a strong pain in the lower abdominal region. He asked Helen Dukas not to call his doctors. The worried secretary, however, secretly instructed Margot, who was in the hospital with her own problems, to alert Dr. Dean, Einstein's physician. On the afternoon of April 13, after he collapsed, Dukas had no choice but to call the doctor. He came quickly and administered a shot of morphine but was reluctant to transfer Einstein to the hospital because he thought the small internal hemorrhage he suspected caused the problem might soon reabsorb. More specialists arrived the next day, however, and Einstein became irritated with them—he wanted them to leave him alone—though he tried to remain good-natured. All of them, including Einstein, now knew that the aneurysm had ruptured and death was imminent. Einstein only asked the doctors if it would be a "horrible death." They said they didn't know. Helen Dukas fussed over him. "You're really hysterical—I have to pass on sometime, and it doesn't really matter when," he scolded her. After his condition worsened over the next two days, his doctors insisted that he enter Princeton Hospital so he could be fed intravenously and the nurses could regularly administer shots of morphine to deaden the pain. The hospital staff tried to make him comfortable.

During the night of April 17–18, Einstein slept peacefully. Soon after midnight, he died in his sleep. Alberta Roszel, the night nurse, was the last person to see him alive. She reported, "He gave two breaths and expired." His stepdaughter Margot described his last hours in a letter the same month to a family friend, Hedwig Born: "He waited for his end as for an impending natural event. He faced death quietly and modestly, and was fearless as he had been in life. He left this world without sentimentality and without regrets."

Chapter 12

EINSTEIN THE EXPERIMENTER

> In my old age, I am acquiring a passion for experiment.
>
> —Albert Einstein, 1915, age 36

Einstein was interested in far more than only abstract things. It is true that special and general relativity fly in the face of everyday experience, but Einstein was also fascinated by practical, down-to-earth phenomena. In his physics, he usually sought out ways in which his theories could be tested and, it was hoped, confirmed by experiment. Sometimes, as with his early work on molecules, Einstein took an everyday phenomenon and looked at it in an entirely new way. Then he gained some insight into the nature of things that could have been seen by anyone else—if only they had looked.

Take the papers of 1905, for example. His Ph.D. thesis showed how to find out Avogadro's number. By plugging in real data for a sugar solution, Einstein got his own value. In Brownian motion, he explained the odd movement of pollen grains that had been reported about 75 years earlier. In 1906, Einstein extended these results to show how particles are distributed in a tall column of liquid, something relatively easy for experimenters to check. He also wrote papers to explain Brownian motion to chemists who may not have had the mathematical training expected of a physicist. His papers on special relativity make mention of ways they can be tested via length contraction, time dilation, and radioactive decay. This is not the behavior of an abstract theoretician.

While Einstein was interested mostly in quantum physics, statistical physics, and gravitation, he still found time to explore other areas of physics. Though it is not well known, Einstein wrote an article, "On the Cause of Meanders in the Courses of Rivers," which appeared in the journal *Naturwissenschaften* (Natural Sciences) in 1926. At that time, quantum physics was in its infancy, and Einstein was deeply involved in its development. Nevertheless, he took time away from quantum theory to explain the strange undulating path that rivers may have when they approach an estuary. He wrote an article on how sound travels through a gas, which may have been a refreshing break from the mathematical rigors of general relativity.

* * *

Einstein thought up and devised many experiments that he did not carry out himself but nevertheless urged others to do. He came up with a new experiment based on the Stern-Gerlach experiment. In the original experiment, silver atoms traveled through a magnetic field. Atomic particles, such as the electron, possess a quantity known as *spin*. This intrinsic spin is not as obvious as it may sound but is related to the magnetic moment of an atom. The magnetic moment of the silver atom is due to one, single electron. After passing through the magnetic field, silver atoms emerge either with "spin up" or "spin down," which allows the spin to be determined. The spin is related to the magnetic moment of the electron, which means the electron acts rather like a tiny magnet. Einstein, fascinated by the experiment, proposed a modification of it. He also proposed a new experiment to explore superconductivity. A Dutch friend, Nobelist Heike Kamerlingh Onnes, was famous for having discovered superconductivity. At low temperatures, some materials lose all resistance to the flow of electricity. If a current starts to flow in a superconductor, it will never diminish. Today, the search is on to find usable room-temperature superconductors that would decrease significantly the amount of money we are spending to produce electrical energy. In 1922, Einstein, in an essay to honor Kamerlingh Onnes, suggested a different way of experimentation to explore the phenomenon. He wrote, "With our wide-ranging ignorance of the quantum mechanics of composite systems, we are far from able to compose a theory out of these vague ideas. We can only rely on experiment." Einstein's essay also included the phrase "quantum mechanics," probably the first time it ever appeared in print.

Einstein was no mere armchair experimenter. Another Dutch friend, Hendrik A. Lorentz, had a son-in-law, Wander de Haas, who was also a physicist. When Einstein was in Berlin, de Haas needed work and came to join Einstein as an assistant in April 1915. Together they worked on the properties of a magnetic field, about which Einstein wrote, "I have done a wonderful experimental thing this semester, together with Lorentz's son-in-law." The paper, "Experimental Proof of Ampere's Molecular Currents," was published the following year, right in the middle of Einstein's major work on general relativity, in the *Proceedings of the German Physical Society*. Later, in 1916, de Haas and Einstein joined forces on another paper, with the similar title, "Experimental Proof of the Existence of Ampere's Molecular Currents." The second article was necessary because Lorentz had spotted a mistake in the paper written by his friend and his son-in-law. The new paper, correcting the error, appeared in a Dutch journal, the *Proceedings of the Royal Academy of Sciences of Amsterdam*.

Ampere suggested, in 1820, that magnetism is caused by electric currents in motion. By 1915, the electron had long since been discovered, and it seemed natural to explore Ampere's idea further. Einstein and de Haas took a large iron cylinder and wrapped a coil around it. They passed an alternating current through the coil, and the changing current altered the cylinder's magnetization. This changing magnetization set up a torque on the cylinder, which began to rotate. To help visualize Einstein and de Haas's work, imagine a group of rowers in a boat. If they all pull randomly, they go nowhere. With a coxswain, who gives commands to keep the rowers paddling together, the efforts add up, and the boat moves swiftly through the water. Einstein and de Haas thought that magnetization acts like a coxswain and gets all the electrons orbiting the atoms in the iron cylinder to move together. When the angular momentum of the electrons is in sync, the cylinder starts to rotate. Einstein and de Haas described theoretically the motion of an electron orbiting in a circular path. They found a relation between the magnetization and the angle through which the cylinder rotates. Their idea, that a magnet can behave like a gyroscope, is now known as the *Einstein–de Haas effect*. Einstein wrote that "the experiment yielded in all detail a confirmation of the theory." The two scientists were not right, but at the time their answer seemed correct. Niels Bohr, who applied the idea of the quantum to the atom, loved the Einstein–de Haas effect. Bohr was criticized for his model of the atom, in which electrons moved in circles around the nucleus. Traditional physics said this could never happen. An electron moving in a circle should radiate energy and quickly spiral into the center of the atom. Einstein and de Haas provided experimental support for Bohr's idea. Bohr wrote, "As pointed out by Einstein and de Haas, their experiments indicate very strongly that electrons can rotate in atoms without the emission of energy radiation." It was another important step forward for quantum theory.

If experiments were not enough, Einstein also held several patents. From his days in the Swiss Patent Office, he obviously knew the procedure for obtaining a patent quite well and must have developed a sense of what was patentable and what wasn't. His research on Brownian motion had led Einstein to predict that there should be voltage fluctuations inside a capacitor. He published an article on it with the long title "On the Limit of Validity of the Law of Thermodynamic Equilibrium and on the Possibility of a New Determination of the Elementary Quanta" in *Annalen der Physik* in 1907. To verify his prediction experimentally, Einstein needed a device that could generate and detect tiny fractions of a volt. Einstein, in association with the Habicht brothers, designed it and had it built. Perhaps Hermann Einstein's electrical business had sparked his son's creativity. As with many potential patents, though, there was little or no interest from manufacturers.

Einstein eventually held several patents. Most of them date from the late 1920s and early 1930s, when he joined forces with Hungarian inventor and physicist Leo Szilard. In November 1927, the two scientists applied for a patent on a home refrigerator using what has come to be known as the Einstein-Szilard pump and protected the fundamental principles of the pump in a stream of patents filed in several different European countries, including Britain, Germany, Austria, and Switzerland. The refrigerators of the time were noisy and unreliable, and, worse, they often poisoned their owners through the leaking of toxic refrigerants. The two men proposed a nonmechanical, noiseless absorption refrigerator and then sold this idea and a later one to a division of the Electrolux Company. AEG, the German division of General Electric, bought a third design. In all, the two men collaborated on five designs. But by 1932, the development of these refrigerators was abandoned because of bad economic times and the invention of Freon, a safer coolant. Still, the two scientists continued their seven-year collaboration to make and patent a number of other mechanical gadgets and made some money from the patents. Their design became important once atomic energy became possible, for it was a great way to cool off some of the nuclear material involved in the reactions. Rudolf Goldschmidt registered a joint patent for a loudspeaker with Einstein in 1933, shortly after he had arrived in the United States. The two inventors planned to make a hearing device for a mutual friend, musician Olga Eisner. The patent, filed in Nazi Germany, gave Einstein's address as "whereabouts unknown."

Einstein also helped to develop a gyrocompass with Hermann Kämpfe, and enlisted his young sons' help in doing so. When a quarterback throws a football, he puts a spin on it. This spin keeps the ball pointing in the direction in which it's thrown, which should help the quarterback reach his receiver. Einstein realized that the same principle could serve as a compass. Set something spinning in a "north" direction, and it should stay pointing that way. The idea of the magnetic gyrocompass may have occurred to him while working with de Haas. Kämpfe patented the device, but Einstein received a royalty of 1 percent. Einstein also shared a U.S. patent with Gustav Bucky, a German-born physician and longtime friend, for a self-adjusting light-intensity camera, the forerunner of today's automatic flash cameras. This device makes use of photoelectricity.

* * *

It is a measure of Einstein's greatness that, if you discard the physics for which he is best known, he would still remain one of the greatest, if not the greatest, physicist of all time. During their university studies, physics students encounter special and general relativity, Brownian motion, and the photoelectric effect. They also meet with Einstein's explanations of specific heat and his work on radiation and the statistics of particles. Each of these topics was profound and still generates considerable new research today.

Take a lump of metal and heat it up. How much energy do you have to supply to raise one gram of the metal by one degree? This is the specific heat capacity, c, of the metal. Two French physicists, Pierre Louis Dulong and Alexis Thérèse Petit, discovered, in 1819, that many solid metals have the same specific heat capacity, about 6 cal/mol. This became known as the *Dulong and Petit law,* which states that all substances have the same value for c. By 1840, it was clear that there was a problem, for room-temperature diamond clearly had values of c that were lower than the Dulong-Petit law. Gases fared even worse.

Einstein's former professor in Zurich, Heinrich Weber, made great strides with this problem. He showed that the specific heat capacity of diamond was not constant but depended greatly on temperature. For diamond, between 0 and 200 degrees centigrade, Weber found that c changed by a factor of three. James Dewar, inventor of the thermos, managed to liquefy helium in 1898. This opened up a vast new arena of science, low-temperature physics. Dewar turned his attention, naturally enough, to diamond and measured the specific heat capacity between 20 and 85 Kelvin. He found a value of c of about 0.05 cal/mol, some 100 times less than the Dulong-Petit prediction.

In 1896, Ludwig Boltzmann, a pioneer of statistical physics, made some progress theoretically. Boltzmann, like Einstein's friends Paul Ehrenfest and Max Planck, led a tragic life, but all discovered some of the most important physics of the twentieth century. Boltzmann treated atoms in the solid metal as though they were oscillators, moving back and forth with a definite frequency. By adding up the energy of all the oscillators, he came up with an equation for c that gave the rough answer of $c = 6$ cal/mol. He had thus given some theoretical backing for the Dulong-Petit law. Enter Einstein in 1907. Einstein blended Boltzmann's model with the probabilities of Planck. Planck, in his classic 1900 paper on radiation, proposed that an oscillator has a definite energy but that not all energies are equally probable. Higher-energy oscillators are far more rare than low-energy oscillators. In other words, there is a certain distribution of oscillator energies. Like Boltzmann, Einstein modeled the solid as a collection of oscillators but used the same energy distribution as Planck. The only energies that Einstein allowed were multiples of hf. He added up the energies of all these oscillators. The result was Einstein's law for the specific heats of solids. In his paper, published in 1907, Einstein included a graph showing how his theory explained the data obtained by Weber. At temperatures close to absolute zero, Einstein's expression has to be slightly modified. Peter Debye, in 1912, improved Einstein's formula and explained beautifully the experimentally measured values. To Einstein, though, belonged the honor of being the first to apply quantum theory to a solid. Today, vast textbooks are written on solid-state physics. The advances made in the subject have spawned the computer, the video, and all other devices based on silicon or germanium.

A standard question from children has always been, "Why is the sky blue?" The answer perplexed physicists for many years. Marian Ritter von Smolan Smoluchowski, whose work on Brownian motion had greatly complemented Einstein's, thought he had an answer. In Brownian motion, fluctuations in position are related to time. Smoluchowski analyzed the perfect gas and showed how fluctuations in density are related to temperature. While a gas might seem to be smooth and have a constant density everywhere, Smoluchowski said there will be high- or low-density regions. These clumps occur at random, so you can't predict when or where they will occur, only that they do occur. You can, as Smoluchowski showed, work out how large the fluctuations will be and what they depend on. He showed that at the critical point of the gas, density fluctuations become incredibly large. The critical point is the temperature, at a given pressure, at which both the gas and the liquid can exist. Steam and water at 100 degrees centigrade is an example of a substance at a critical point.

If you place a colored straw in a glass of water, you'll see something odd. The straw will look bent. This optical illusion is caused by the change in the *refractive index* between air and water. The *refractive index* is a measure of the speed of light in a certain material. The speed of light in air is greater than that in water. The net result is that a beam of light headed into water will appear to bend. The same physics shows the problem of "apparent depth." If you look down into a container of water, it appears to contain less liquid than it actually does. For that reason, if you ever have to dispense a liquid medicine, hold the cup up to eye level to make sure the medicine is really at the level it ought to be.

The refractive index of a gas depends on its density, something formally known as the *Gladstone-Dale law*. Smoluchowski thought that density fluctuations caused fluctuations in the refractive index of the gas, so, at the critical point, there would be huge scattering of light by the gas. Smoluchowski had some experiments to back him up. Richard Avenarius had shown, in 1874, that there was strong scattering by a gas close to its critical point. This is *critical opalescence*. John Tyndall, in 1869, suggested that the sky is blue because of the strong scattering of light by droplets of water in the atmosphere, the same droplets responsible for rainbows on a wet yet sunny day. He stated that particles of dust may cause scattering of light, too. For this reason, the trail of smoke from a cigarette is usually blue, while sunsets in polluted skies are often a glorious red.

Einstein refined the work of Smoluchowski and derived a formula that went farther than mere speculation. Einstein's formula predicts the amount of scattering of light of a definite frequency by a gas at a certain temperature at a particular angle. This work was typical Einstein. He had converted some experimental results and theoretical speculations into a rigorous theory. When simplified, Einstein's expression agreed with Lord Rayleigh's approximation of years earlier. As an added bonus, Einstein's formula suggested a new way to determine Avogadro's number. Smoluchowski was so impressed by Einstein's work that he tried to set up a laboratory experiment to test it. The preliminary results looked good, but sadly Smoluchowski passed away before the full set of results was in. Atmospheric scientists would eventually use Einstein's equation to measure Avogadro's number. It was another success.

* * *

Einstein repeatedly turned his attention to the properties of light. Whether it was to explore the light-quantum hypothesis or to look at scattering by a gas, it seems as though radiation was never too far from his mind. In November 1916, he wrote, "A splendid light has dawned on me about the adsorption and emission of radiation." This thought led him to publish three articles, two in 1916 and one in 1917. Einstein began simply enough. He considered a gas of particles immersed in a bath of radiation.

To simplify what Einstein did, suppose the particles can have only one of two possible energies—one low, one high. He asked how many particles will go from high to low energy in a certain time period and how many from low to high. A particle can jump to high energy by absorbing radiation of frequency *f*. They can decrease their energy by emitting radiation of frequency *f*. He introduced two coefficients, now known as Einstein's *A* and *B* coefficients. The first, *A*, described the spontaneous transitions that occur even when there is no radiation. Einstein's *A* crops up only in the equation for jumps from high to low energy. That's only natural, for normally nothing spontaneously increases its energy. The other number, the *B* coefficient, models the transitions induced by radiation. This occurs in both upward and downward transitions.

As an illustration, suppose there is a small wall with some children standing on it. The wall is safe enough to jump from but too high to climb. There's also an elevator that takes children on the ground up to wall level. In the absence of the elevator (radiation), kids on the wall may jump down to the ground. This is the process modeled by the *A* coefficient. Kids transported up to or down from the wall by the elevator are modeled by the *B* coefficient.

If everything is settled and in equilibrium, the number of particles making a downward jump in energy equals the number making an upward jump. This gave Einstein the equation he needed. For high temperatures, his equation had to be the same as Wien's radiation law. If that were the case, Einstein showed, then the particles can absorb or emit radiation of only a particle amount—energies that are multiples of *hf*. He had provided yet more evidence to support the light-quantum hypothesis.

An added bonus was hidden in Einstein's theory. If particles can be "pumped" to a higher level, then as they drop back down to a lower energy level, they give off energy in the form of radiation. All this radiation has the same frequency. This is the principle of a laser, which stands for "light amplification by the stimulated emission of radiation." A short burst of light pumps some of the atoms in the laser to a high-energy level, and they then drop down to a lower level, producing laser light as they do so. It's rather like the elevator "pumping" children on to the wall, and then, as they jump off, all of them make an identical "thud" as they land. The developers of the laser, Arthur Schawlow and Charles Townes, would earn the Nobel Prize for their device, one of a number of laureates who built on Einstein's work.

If Einstein's way of counting transitions between energy levels worked well, his way of counting particles paid out in spades. In June 1924, Einstein received a letter, in English, from a young Indian physicist, Satyendra nath Bose. The young man had just received a rejection for an article from *The Philosophical Magazine* but did not understand why. Bose asked Einstein to take a look at the paper to see if anything was wrong. If not, he asked, would Einstein help him get it published in *Zeitschrift für Physik* (Journal of Physics)? Einstein confirmed Bose's results and helped the physicist get the paper published, though he first had to translate it into German. It would turn out to be an influential paper, though Bose had the honesty to admit later on that "I had no idea that what I had done was truly novel." He had set Planck's radiation law on a far firmer footing.

Think of a collection of particles that do not possess mass. What's more, assume they exist in one of two polarizations (say, up and down). Further, the number of these particles is not fixed in time. Bose broke up space into a group of boxes of a certain volume. He asked in how many "states"—in how many ways—particles could be in each box, with momentum in a certain small range. This method of counting was, in 1924, considered extremely odd. He counted not particles but "states," but no one knew quite what to think of them. Bose, though, made one last step. He assumed that the momentum in these states was hf/c. With this final masterstroke, his work was done. Out of all the numbers and Bose's weird counting scheme fell Planck's formula for radiation. It implied, yet again, that radiation behaved like massless particles with energy hf and momentum hf/c.

Einstein took Bose's idea a step farther. He combined this new method of counting and the parallel with photons to come up with something entirely new: the quantum gas. It was a new form of particles. They could have mass, but the number of them in any given cell with a certain momentum is given exactly by Bose's rule of counting. Particles that can be counted this way are called *bosons*, in honor of the Indian physicist. Einstein cleaned up Bose's derivation and made a bold prediction. Particles that can be counted this way, which obey "Bose-Einstein statistics," have strange low-temperature behavior. As the Bose gas drops below a certain temperature, Einstein predicted that "a saturation is effected; one part condenses, the rest remains a saturated ideal gas." This is called *Bose-Einstein condensation*, in which all the bosons in the gas occupy precisely the same state. The Bose-Einstein condensate is a type of matter that is distinctly different from other things on Earth. In December that year, Einstein wrote to Paul Ehrenfest, "The theory is pretty, but is there some truth to it?" In the 1990s, physicists working at the National Institute for Standards and Technology in Boulder, Colorado, succeeded, finally, in making a Bose-Einstein condensate. The experimental challenges they overcame were so immense that the scientists were awarded the Nobel Prize in 2001.

Shortly after the Bose-Einstein work of 1925, physicists found a different way to count particles. In that year, Wolfgang Pauli came up with the *Pauli exclusion principle*, which said that no two electrons could be in the same state. In terms of Bose's boxes, it meant there could be only two particles per box—one with spin up, another with spin down. Italian American physicist Enrico Fermi and British physicist Paul Adrien Maurice Dirac followed up on this idea. They proposed that electrons were only one instance of an entire class of particles. These particles, which became known as *fermions*, were to be counted differently from bosons. Fermions, they proposed, all had the same properties as electrons, so, following Pauli's exclusion principle, only two fermions can occupy the same state at the same time. This led to a new method of counting—Fermi-Dirac statistics. Einstein may not have been impressed. He said that he "did not understand Dirac at all" and that "this balancing on the dizzying path between genius and madness is awful."

The characteristic difference between bosons and fermions is their intrinsic spin. Bosons, such as light, have integer spin: 0, 1, 2 … Fermions, such as electrons, have half-integer spins. In some circumstances, fermions can get together to produce bosons. One example is in superconductivity. Leon Cooper showed that electrons, which are fermions, can join together to form so-called *Cooper pairs*. These Cooper pairs behave like bosons, and, as Einstein put it, "a saturation is effected." For this theoretical explanation of superconductivity, Cooper would win the Nobel Prize together with John Bardeen and Robert Schrieffer.

* * *

Einstein's last significant paper was published in 1935 in the American journal *Physical Review*, but he was not its only author. It was one of the few papers Einstein cowrote that has made a lasting impact. Titled with the question "Can Quantum-Mechanical Description Be Considered Complete?," its answer was a resounding no. Einstein had long been dissatisfied with quantum mechanics and had taken part in many debates on the subject. Here, with young Nathan Rosen and Boris Podolsky, he put the cat among the pigeons. In Einstein's own words, the paper "created a stir among physicists and played a large role in philosophical discussion."

It is odd that Einstein, who seized hold of the light-quantum hypothesis to explain the photoelectric effect, who calculated the momentum of a photon, and who routinely treated light as particles was ill at ease with quantum mechanics. Yet he was. In a letter to his friend Heinrich Zangger in Zurich in May 1912, he famously remarked, "The more success the quantum theory has, the sillier it looks," just after Einstein had striking success in calculating specific heats. In 1924, as he was working on the Bose-Einstein statistics, he wrote to Paul Ehrenfest, "The more one chases after quanta, the more they hide themselves."

Max Born, in 1926, interpreted quantum mechanics in terms of probabilities. His view, shared by others, was that the best that physics could achieve was a list of probable outcomes. In quantum mechanics, as particles behave like waves, the equations of quantum mechanics say we can tell only where a particle will *probably* be and what momentum it will *probably* have. We can do no better than that. In early 1926, Einstein wrote to Born, "The Heisenberg-Born concepts leave us all breathless and have made a deep impression on all theoretically oriented people." Later that year, after Born had published papers containing his probabilistic description of quantum mechanics, all bets were off. Einstein told him that "quantum mechanics is very worthy of regard. ... The theory yields much, but it hardly brings us close to the secrets of the Old One. In any case, I am convinced that God does not play dice." Einstein wanted certainty, not probabilities.

Einstein was puzzled by the lack of distinction between particles and waves. What *are* these waves? What do they mean in terms of physics? Einstein thought they might be interpreted as pilot waves, guiding the particle along. He sought a description in which such "wave-particle" duality goes away, *thinking that* some new physics would emerge to reconcile both aspects of behavior. Things became worse when Werner Heisenberg introduced his uncertainty principle in 1927. He said that you cannot know exactly both the position and the momentum of a particle. There would always be some uncertainty in the measurement. The same mathematics shows that for a note on a piano to be perfectly pure, it has to last forever. There's a trade-off between the quality of the pitch and the length of the note. In quantum mechanics, there's a trade-off between knowing where a particle is and what momentum it possesses. More important, Heisenberg's principle did away with the notion of causality. Causality says that an event cannot occur until something *causes* it to take place. A pot of water, for example, boils only *after* you heat it up. Einstein thought that abandoning causality was too high a price to pay for the new quantum mechanics. Leading the charge against causality was Niels Bohr. Einstein wrote to Bohr, "Not often in my life has a human being caused me such joy by his mere presence as you have done." Einstein would later describe Bohr as a "man of true genius," but the two of them disagreed deeply on what the new quantum mechanics meant. A series of debates between the two did not resolve the dispute. In debate, Einstein was gifted at creating "thought experiments" that revealed problems in Bohr's interpretation. One such thought experiment was the Einstein-Podolsky-Rosen paper of 1935, which gave birth to the EPR paradox.

Nathan Rosen was a Brooklynite, while Boris Podolsky was a Russian émigré to the United States. The two young men were at the Institute for Advanced Study in 1934–1935, where Einstein also had only recently arrived. The three men collaborated on the EPR paradox, which was originally thought up by Rosen but written down by Podolsky. The idea is to use quantum mechanics to argue against itself. The paradox then has to be unraveled by those who favor the Bohr description. The experiment at the heart of the EPR paradox is simple. Produce two particles together so that their overall momentum and position are known. Send them far away from each other—so far that information cannot be "exchanged" between them. Then, if you measure the spin of one particle, you instantly can infer the spin of the other particle because the spin of the *pair* of particles were known right at the start of the experiment. This would be fine, except that the laws of quantum mechanics say that you can't know precisely the momentum and position of the second particle. By analogy, suppose I always wear a red sock and a blue sock. In terms of a statistical probability, there's a 50 percent chance that my left sock is red. Bohr would say that's the most we can know. If you now enter a sock-hurling machine, which rips the sock off your right foot and hurls it far away, someone will find your sock. If it is red, they know with certainty, without ever having seen you, that you are wearing a blue sock on your left foot. So, if we can only hope to have a statistical account of your sock-wearing habits, the sock-hurling machine shows that it is not complete. Einstein, who never wore socks, would say this shows we *can* know more than mere probabilities. Einstein, Podolsky, and Rosen devised an intellectual equivalent of a sock-hurling machine.

For quantum mechanics, this creates a conundrum: If quantum mechanics is true, then it can't be complete. And if quantum mechanics is not complete, then what theory can we discover that is complete? Can we come up with a theory that does not rely on statistical interpretations and does away with causality but restores complete knowledge and resurrects causality?

The EPR paradox spawned a whole industry. Many different interpretations of quantum mechanics have been proposed, but the EPR paradox is a stiff test for many of them. Princeton physicist David Bohm, who ran afoul of the House Un-American Activities Committee and was arrested for refusing to testify but subsequently acquitted, tried a different approach. Similar in spirit to Einstein, who tried to hire him after Princeton University let him go, Bohm suggested that so-called hidden variables might exist. These, when added into quantum mechanics, might provide a coherent picture. Hidden variables served in many respects like Einstein's pilot waves, but they suffered a severe setback in 1964 by the work of John Bell, who many view as having ruled them out altogether. Philosophers of science, though, are weary of such discussions: working out what the meaning of a single particle and its behavior might be is no longer of interest to them. The work of Bose, Dirac, Einstein, and Fermi on collections of particles have shown philosophers the way. In the philosophy of physics, new emphasis is being put on many-particle systems and how to count them. Instead of Bohr or Einstein, the new philosophical debate is on how or whether elementary particles can be counted at all, if they are supposedly indistinguishable. Imagine going to a party where all the other guests are clones of a single person and they keep moving about. How could you count how many guests there are?

It is a testament to Einstein that physicists, historians of physics, and philosophers of physics—all in their own way—spend a huge amount of time trying to unravel the mysteries of quantum mechanics and general relativity, two fields that he helped found a century ago. He left us with a continuing search for his "theory of everything."

EPILOGUE: EINSTEIN'S LEGACY

Einstein's life, as we have seen, was one of paradoxes. Although he called himself a "dedicated pacifist," he urged President Franklin D. Roosevelt to begin a nuclear weapons program for fear that Germany would develop an atomic bomb during World War II. A humanist and lover of children and animals, he often neglected his own sons and wives. A self-proclaimed loner, he enjoyed many deep friendships, conducted an enormous correspondence, and supported organizations and causes to which he dedicated himself. Proud of his Jewish roots and a cultural Zionist, he did not adhere to Jewish traditions or follow the Jewish faith. A nonbeliever in conventional religious doctrines and a personal God, he held a deep faith in the laws and harmony of nature that he called "cosmic religion." As a young man, he detested Germany for its authoritarian and rigid institutions, yet he returned there to work for many years and enjoyed the benefits of its scientific community before the rise of Hitler. He remained modest and humble throughout his life, attributing his achievements to a childish curiosity and a willingness to flout conventional wisdom. Although he was publicity-shy and reclusive, Einstein skillfully used the press to his advantage in interviews and used his fame to pursue the causes in which he believed. The physics he established in the early part of the twentieth century has led to technological and scientific advances unprecedented in human history.

Appendix

THE STORY OF EINSTEIN'S FAMOUS BRAIN

Einstein's body, from which the brain and eyes were removed during the autopsy, was cremated late in the afternoon of April 18, the day he died. His ashes were scattered, probably over the Delaware River not far from Princeton, by his friends Otto Nathan and Paul Oppenheim. The news spread rapidly throughout the world as a flood of tributes filled the media.

What happened to Einstein's body shortly after his death has become well known. His brain and eyes were preserved for future study. The pathologist, Dr. Thomas Harvey, who performed an autopsy, removed the brain without permission and kept it in a jar of formaldehyde. Another pathologist, Dr. Henry Abrams, took the eyes with the permission of the hospital administrator and received a letter of authenticity from Dr. Guy Dean, Einstein's personal physician at the time of his death. Apparently this procedure is not uncommon during autopsies.

After the cremation, the astounded Einstein family learned about the removal of the brain. According to the account of Carolyn Abraham in her book *Possessing Genius*, Otto Nathan was in the morgue during the autopsy and aware of the removal. The family agreed to let Dr. Harvey keep the organ if he used it only for scientific study. He gave at least three parts of it to other scientists, but until recently only one of them had made any use of it. In 1985, Professor Marian Diamond of the University of California at Berkeley, in an article in the medical journal *Experimental Neurology*, reported that Einstein's brain had an above-average number of glial cells (which nourish neurons) in the areas of the left hemisphere that are thought to control mathematical and linguistic skills. Since then,

Sandra Witelson, a neuroscientist at McMaster University in Ontario, Canada, published some research results on the brain in June 1999 in the British medical journal *Lancet*. Witelson's group conducted the only study of the overall anatomy of Einstein's brain after Dr. Harvey offered to give them a section of it in 1996. The researchers compared Einstein's brain with the preserved brains of 35 men and 56 women known to have normal intelligence when they died. They discovered that in Einstein's case, the part of the brain thought to be related to mathematical reasoning—the inferior parietal lobe—was 15 percent wider than normal on both sides. Furthermore, they found that the Sylvian fissure, the groove that normally runs from the front of the brain to the back, did not extend all the way in Einstein's case. Witelson theorized that this latter feature may be the key to Einstein's intelligence because the absence of a full groove may have allowed more neurons in this area to establish connections among one another and work together more easily. Other parts of Einstein's brain appeared to be a bit smaller than average, putting overall brain size and weight within a normal range.

The odyssey of the brain has been described in books by Michael Paterniti and Carol Abraham (see the bibliography). The brain was recently returned to Princeton Hospital, now the University Medical Center at Princeton, to the care of pathologist, Dr. Elliott Krauss.

BIBLIOGRAPHY

Abraham, Carolyn. *Possessing Genius: The Bizarre Odyssey of Einstein's Brain*. New York: St. Martin's, 2001.

Bernstein, Jeremy. *Quantum Profiles*. Princeton, N.J.: Princeton University Press, 1991.

Calaprice, Alice. *Dear Professor Einstein: Einstein's Letters to and from Children*. Amherst, N.Y.: Prometheus Books, 2002.

———. "Einstein's Last Musings." *Princeton University Library Chronicle*, Autumn 2003.

———. *The Einstein Almanac*. Baltimore: Johns Hopkins University Press, 2004.

———. *The New Quotable Einstein*. Princeton, N.J.: Princeton University Press, 2005.

Clark, Ronald W. *Einstein: The Life and Times*. New York: Crowell, 1971.

Collected Papers of Albert Einstein (CPAE). Various editors and translators. Vols. 1–9. Princeton, N.J.: Princeton University Press, 1986–2004.

Durrell, Clement V. *Readable Relativity*. Foreword by Freeman Dyson. London: Bell and Sons, 1962.

Einstein, Albert. *Ideas and Opinions*. New York: Crown, 1954.

———. *The Meaning of Relativity*. 5th ed. Princeton, N.J.: Princeton University Press, 1956. Princeton Science Library edition (pbk.), 1988.

Fölsing, Albrecht. *Albert Einstein*. New York: Viking, 1997.

Gardner, Howard. *Creating Minds: An Anatomy of Creativity*. New York: Basic Books, 1993.

Highfield, Roger, and Paul Carter. *The Private Lives of Albert Einstein*. London: Faber and Faber, 1993.

Jammer, Max. *Einstein and Religion*. Princeton, N.J.: Princeton University Press, 1999.

Jerome, Fred. *The Einstein File: J. Edgar Hoover's Secret War against the World's Most Famous Scientist*. New York: St. Martin's, 2002.

Kantha, Sachi Sri. *An Einstein Dictionary*. Westport, Conn.: Greenwood Press, 1996.

Kragh, Helge. *Cosmology and Controversy: The Historical Development of Two Theories of the Universe*. Princeton, N.J.: Princeton University Press, 1999.

———. *Quantum Generations: A History of Physics in the Twentieth Century*. Princeton, N.J.: Princeton University Press, 2002.

Overbye, Dennis. *Einstein in Love: A Scientific Romance*. New York: Viking, 2000.

Oxford Dictionary of Physics. 4th ed. Ed. Alan Isaacs. Oxford: Oxford University Press, 2000.

Pais, Abraham. *Subtle Is the Lord ...The Science and the Life of Albert Einstein*. New York: Oxford University Press, 1982.

Paterniti, Michael. *Driving Mr. Albert: A Trip across America with Einstein's Brain*. New York: Dial, 2000.

Popović, Milan. *In Albert's Shadow: The Life and Letters of Mileva Marić, Einstein's First Wife*. Baltimore: Johns Hopkins University Press, 2003.

Princeton University Library Chronicle. Vol. 65, no. 1 (Autumn 2003). A special issue containing articles by Alfred Bush and Alice Calaprice about the Fantova-Einstein relationship and poems that Einstein wrote to Fantova, translated by Alfred Engel.

Renn, Jürgen, and Robert Schulmann, eds. *Albert Einstein/Mileva Marić: The Love Letters*. Trans. Shawn Smith. Princeton, N.J.: Princeton University Press, 1992.

Roboz Einstein, Elizabeth. *Hans Albert Einstein*. Iowa City: University of Iowa.

Rosenkranz, Ze'ev. *The Einstein Scrapbook*. Baltimore: Johns Hopkins University Press, 2002.

Schilpp, Paul, ed. *Albert Einstein: Philosopher-Scientist*. Evanston, Ill.: Library of Living Philosophers, 1949.

Stachel, John, ed., with the assistance of Trevor Lipscombe, Alice Calaprice, and Sam Elworthy. *Einstein's Miraculous Year*. Princeton, N.J.: Princeton University Press, 1998. Reissued 2005.

INDEX

About the Authors

ALICE CALAPRICE was senior editor at Princeton University Press for over 20 years, specializing in editing the sciences. She was in charge of editing and production of the *Collected Papers of Albert Einstein* and was administrator of the series' accompanying translation project. She is the author of several books on Einstein, including *The Quotable Einstein.*

TREVOR LIPSCOMBE was educated at the University of London and then at Oxford, where he obtained a doctorate in theoretical physics. He has published widely in quantum mechanics and statistical physics, two of Einstein's main areas of interest.